Collins

Weather
ALMANAC
A GUIDE TO
2026

Zoë Johnson

Published by Collins
An imprint of HarperCollins Publishers
1 Robroyston Gate,
Glasgow G33 1JN
collins.co.uk

HarperCollins Publishers
Macken House
39/40 Mayor Street Upper, Dublin 1
D01 C9W8, Ireland

First published 2025

© HarperCollins Publishers 2025
Text © Zoë Johnson 2025
Cover illustrations © Julia Murray
Images © see Sources page 266

Collins ® is a registered trademark of HarperCollins Publishers Ltd
All rights reserved. No part of this publication may be reproduced, stored
in a retrieval system, or transmitted, in any form or by any means, electronic,
mechanical, photocopying, recording or otherwise without the prior permission
in writing of the publisher and copyright owners.

Without limiting the exclusive rights of any author, contributor or the publisher
of this publication, any unauthorised use of this publication to train generative artificial
intelligence (AI) technologies is expressly prohibited. HarperCollins also exercise
their rights under Article 4(3) of the Digital Single Market Directive 2019/790
and expressly reserve this publication from the text and data mining exception.

The contents of this publication are believed correct at the time of printing.
Nevertheless the publisher can accept no responsibility for errors or omissions,
changes in the detail given or for any expense of loss thereby caused.

HarperCollins does not warrant that any website mentioned in this title will be provided
uninterrupted, that any website will be error free, that defects will be corrected, or that the
website or the server that makes it available are free of viruses or bugs. For full terms and
conditions please refer to the site terms provided on the website.

A catalogue record for this book is available from the British Library

ISBN 978-0-00-874781-7

10 9 8 7 6 5 4 3 2 1

Printed in the UK using 100% Renewable Electricity at CPI Grou (UK) Ltd

If you would like to comment on any aspect of this book,
please contact us at the above address or online.
Email: collins.reference@harpercollins.co.uk

This book contains FSC™ certified paper and other controlled
sources to ensure responsible forest management.

For more information visit: www.harpercollins.co.uk/green

Contents

4	Introduction
17	**January**
33	**February**
49	**March**
65	**April**
81	**May**
97	**June**
113	**July**
129	**August**
145	**September**
161	**October**
177	**November**
193	**December**
209	Additional Information
211	Our Regional Climates
234	Clouds
248	How to Use Weather Apps
251	The Beaufort Scale
254	The International Tornado Intensity Scale
259	The TORRO Hailstorm Intensity Scale
261	Twilight Diagrams
265	References and Image Sources
267	Acknowledgements
268	Index

Introduction

Our variable weather
In the UK, there's nothing we love more than to talk about the weather and if you've picked up this book, you probably like it more than most. Perhaps it's because the weather here is always changing from one thing to the next. Through the seasons, we experience nearly every weather condition under the sun, from extreme winter storms to baking summer heatwaves. Our position close to the Atlantic Ocean, but also near to continental Europe, allows us to enjoy a range of different air masses. Air travelling over the Atlantic Ocean or Arctic Sea is typically moist, often resulting in wet weather when it reaches us. Meanwhile, air masses arriving from the Continent to the east or southeast are typically much drier and tend to bring extremes of temperature, such as very cold air from the east during winter and hot continental air all the way from North Africa in summer. However, most of our weather comes from the west, over the Atlantic Ocean, and it's this we have to thank for our mild, maritime climate. The transport of warm water northeastwards from the Gulf of Mexico gives the UK much milder weather than other areas of a similar latitude.

Another thing that makes our weather so changeable is the jet stream. This ribbon of strong winds about 8–12 km above the Earth's surface is responsible for steering low-pressure systems, or depressions, towards the UK. It's strongest in winter, due to the difference in temperature between the cold North Pole and warm Equator, and typically located to the south of the UK, with us on the cold side. It brings a constant stream of depressions in from the west, often strengthening them as it does so. In the summer, it's much weaker and usually located to the north of the UK, allowing our weather to become more settled and influenced by areas of high pressure. Occasionally, the jet stream can buckle and become much slower and more meandering, known as meridional. This can lead to a blocking situation, where depressions can no longer move eastwards and the UK becomes 'stuck' under a particular weather pattern.

INTRODUCTION

How to use this book
There is so much to be gained from observing the weather, from watching how it changes through the seasons to appreciating the beauty of a whimsically shaped cloud. The pages that follow will guide you through each month of the year, outlining the gradual shifts in our weather as the seasons move from one to another. For those that love a good fact, extremes of temperature and pressure are provided each month, as well as the average weather that can be expected, based on data gathered by the UK Met Office during the 30 years between 1991 and 2020. For a bit of weather nostalgia, annual and monthly summaries are provided of the weather in 2024, the most recent year in which data is fully available at the time of publishing. Any records broken, such as the warmest or wettest month on record, are based on the Met Office HadUK data set. From 1836 to the present day, historical weather observations are interpolated to give consistent coverage across the UK rather than at just a few points, allowing comparisons to be drawn. Key dates and meteorological phenomena to look out for in each month are given, as well as some weather-related anniversaries. For 2026, a new section has been introduced to each chapter: 'Observing the weather'. Every month, this section includes a simple science experiment or weather event to look out for. From watching ice crystals form before your own eyes to learning how to mark a frontal passage by the changes in the cloud, hopefully you feel encouraged to indulge your inner scientist and try a few.

Tables of sunrise, sunset, moonrise and moonset are also given, as well as the usual twilight diagrams and Moon phases, details of which are explained further on pages 8–9. Finally, each chapter ends with a weather-related story tied to the month it's found in.

At the back of the book, you will find more information about the climate in different parts of the UK, cloud types and measuring the weather.

Hopefully you enjoy making the Weather Almanac part of your year, taking time to note the monthly changes that give British weather its reputation for variety.

The seasons
From the fierce beauty of winter to the glorious warmth of summer, the seasons bring a state of near-constant change to our nature and weather. Every time you step outside, there is something new to marvel over. Green shoots emerging in early spring, spectacular summer flowers, the rustle of falling leaves in autumn and clouds of breath swirling as you exhale into the freezing winter air. The gradual shift from one season to the next is a gift, especially to those who love the weather.

Our four distinct seasons are a result of the Earth's tilt. Our planet rotates on an axis angled at 23.5 degrees. It also orbits the Sun and as it does so, some parts are tilted more towards the Sun, and some away. This changes throughout the year it takes Earth to complete a full orbit and it's what gives us our seasons. In the UK, summer is when the northern hemisphere is tilted towards the Sun. In autumn it begins to move away, until the northern hemisphere is angled away from the Sun in winter. As spring progresses, we begin to tilt back towards the Sun.

Meteorologists define each season as three full months. Spring begins on 1 March, summer on 1 June, autumn on 1 September and winter on 1 December. Keeping to consistent dates allows scientists to make statistical comparisons between the seasons. Meanwhile, the astronomical seasons are based on Earth's position as it rotates around the Sun and each begins with the date of an equinox or solstice. These change slightly every year, making it difficult for statistical calculations and comparisons – hence the invention of meteorological seasons in the late eighteenth century. In 2026, astronomical spring begins at the equinox on 20 March, summer at the 21 June solstice, autumn at the equinox on 23 September and winter starts at the December solstice on the 21st.

Generally, both meteorological and astronomical winter mark the coldest time of year and summer is the warmest, while autumn and spring represent the periods of transition in between. In the UK, autumn and winter are typically the most unsettled seasons as a strong jet stream brings Atlantic depressions in from the west. Spring is most likely to be settled, with an increased likelihood of atmospheric blocking (see page 57) and although warm, or even hot, summer brings the greatest chance of thunderstorms.

INTRODUCTION

Meteorologically and astronomically are the most common ways to split the year into four seasons. However, in 1950, climatologist Professor Hubert Lamb published a scientific paper suggesting five natural seasons. Analysing more than 18,000 daily weather charts, he looked for patterns and trends that appeared each year. This led him to define the following seasons: early winter, from mid-November to mid-January; late winter and early spring, from mid-January to the end of March; spring and early summer, from April to mid-June; high summer, from mid-June to early September; and autumn, from early September to mid-November. Many of his discoveries still apply today; for instance, the increased likelihood of blocking and easterly winds in spring and early summer. However, his five seasons were based on historical data, the most recent of these nearly 80 years old. While ground-breaking at the time, our climate has changed significantly in the past 100 years and trends that were observed back then may have also changed.

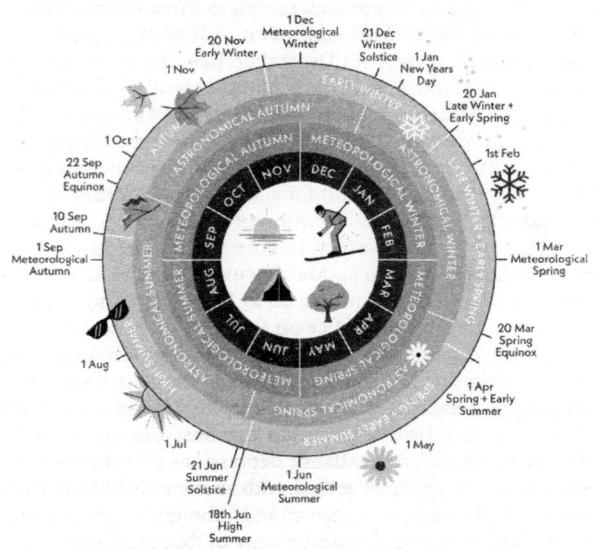

Meanwhile, some ecologists divide the year into six seasons based on changes in nature, such as flowers coming into bloom and animals going into hibernation. These are: prevernal, vernal, estival, serotinal, autumnal and hibernal. Elsewhere around the world, many cultures also divide the year into six, while some only experience two seasons, a wet and a dry.

Sunrise and sunset
Intrinsically linked to the seasons is the length of day and night, which is also based on the Earth's tilt. While the northern hemisphere is angled towards the Sun in summer, the UK basks in 16 to 18 hours of daylight, depending on latitude, each day. Areas at a higher latitude in the north receive more daylight in summer and less in winter, as they are closest to the North Pole. In winter, a scarce six and a half hours of sunlight illuminates northern Scotland on the solstice, while southern England receives eight hours of daylight.

For four dates each month, tables show the times of sunrise and sunset at the four capital cities in the United Kingdom: Belfast, Cardiff, Edinburgh and London. Times are given in Coordinated Universal Time (UTC) during the winter half-year, also known as Greenwich Mean Time (GMT), but are adjusted to account for daylight savings in British Summer Time (BST). The times shown also take the effect of refraction into account, whereby light bending as it enters the atmosphere makes the Sun and Moon appear visible a few minutes before they have risen above the horizon, and a few minutes after they have set.

The Moon and its phases
More than just a celestial body illuminating the night sky, the Moon is incredibly important to life on Earth. It not only causes tides with its gravitational pull, but also slows the Earth's rotation, giving us our 24-hour days. The Moon orbits Earth once every 27.3 days, simultaneously completing one turn on its axis. This means we only ever see one side of it. As it completes its orbit, the amount of the Moon illuminated by the Sun appears to change to a viewer on Earth. This cycle of waxing and waning from New Moon to

INTRODUCTION

Full Moon and back takes about 29.5 days. For each month, a table showing the Moon's phases, together with the age of the Moon counted from New Moon is given. The times given for moonrise and moonset may appear in either order, depending on whether the Moon rises in the morning and sets in the evening, or vice versa. This corresponds with the Moon's phase: when the Moon is at its fullest it reaches its highest point in the sky during the night. Meanwhile a New Moon reaches the highest point during the day, so on the whole, we can't see it. On rare occasions, no time is given for an event and this is because there can be more than 24 hours between successive moonrises or moonsets.

Historically, many cultures have named the Full Moons as a way of marking the passage of time. Many of those used to this day were first named by Native American people, reflecting the changes in the natural world taking place each month. Each month, the most popular name is given, as well as a few alternatives from other cultures.

Twilight

When the Sun has set and night has not yet fallen, there is a magical time of day illuminated by a faint glow in the sky, known as twilight. In the evening, this is also called dusk, while the period before the sun has fully risen in the morning is known as dawn. Like the times of sunrise and sunset, the timings of twilight also vary significantly across the UK. In each month, twilight diagrams are included for the four capital cities, while charts for the full year are included at the back (on page 259–262). There are three stages to twilight, depending on how far the Sun has moved below the horizon. Civil twilight comes first, when the centre of the Sun is less than 6 degrees below the horizon. It's followed by nautical twilight, when the Sun is between 6 and 12 degrees below the horizon. The brightest stars begin to appear, but the horizon is still visible; this allowed sailors to navigate, hence the name of this phase. The final stage is astronomical twilight, when the Sun is between 12 and 18 degrees below the horizon. After that, the night descends into full darkness.

The Weather in 2024

The year 2024 will go down in the record books as warm, wet and full of impactful weather. There were nine named storms, bringing a plethora of severe weather, from extensive flooding to strong winds. May was the warmest on record and also contributed to the UK's warmest spring on record. The second-warmest February and fifth-warmest December meant mild nights and far fewer frosts than expected.

As the New Year was born, 2024 started as it meant to go on: wet. Storm Henk brought widespread flooding to central and southern England and Wales. Soon after was one of the few notable cold spells of the year, before the end of the month returned to unsettled weather, with the powerful Storm Isha bringing a gust of 99 mph to Northumberland. It was the strongest storm since Storm Eunice in 2022. February was exceptionally wet in the south, with both southern England and East Anglia experiencing their wettest February on record. Spring didn't bring any respite until later in the season, as March and April continued in an unsettled vein. In May, temperatures soared into the high 20s, well above average for the time of year. Usually characterized by fine, sunny weather, atmospheric blocking and northeasterly winds, the spring of 2024 bucked the trend, coming out exceptionally warm, wet and dull. In contrast, summer 2024 was unusually cool, with the lowest average temperatures since 2015. There were only occasional short-lived periods of warmth, one of these bringing the hottest day of the year in mid-August. This month saw the final named storm of the 2023/24 season, Storm Lilian. It brought strong winds, heavy rain and a dramatic end to the season; the first time the letter 'L' had been reached since alphabetical naming began in 2015. Autumn and early winter saw significant periods of flooding. Bedfordshire and Oxfordshire saw more than 300 per cent of their average rainfall in September, making it their wettest calendar month on record. In late autumn, Storm Bert brought another round of flooding to southern England and Wales and the wettest calendar day since 2020, with some places recording more than 150 mm of rain. In December, Storm Darragh brought strong winds comparable to Storm Isha and after a relatively settled Christmas period, further unsettled weather brought the year to a close.

INTRODUCTION

Although the summer was cool and there were a few notable cold spells in mid-January and mid-November, it was still the fourth-warmest year on record for the UK, with eight out of 12 months warmer than average. With annual mean temperatures of 9.78 °C, 0.64 °C above the long-term average, only 2022, 2023 and 2014 have been warmer.

The Global Climate in 2024

Globally, 2024 was the warmest year on record and the first time global temperatures exceeded 1.5 °C above pre-industrial levels – the limit set in the Paris Agreement (see pages 62–63). Torrential rainfall brought catastrophic flooding to eastern Spain, an unusually active Atlantic hurricane season saw the deadliest hurricane in the continental US since Katrina in 2005 (Hurricane Helene), while Super Typhoon Yagi brought widespread destruction to Vietnam. Elsewhere, severe drought, heatwaves, wildfires and flooding affected people around the world. The record-breaking warmth of 2024 can partly be attributed to a powerful El Niño event, ending in April. This sees weaker winds in the eastern Pacific, resulting in warm water staying closer to the ocean surface for longer. These events raise global temperatures by about 0.2 °C on average and the 2023/24 El Niño was one of the five strongest on record. However, human-induced climate change is responsible for most of this change. The burning of fossil fuels releases greenhouse gases, such as carbon dioxide (CO_2), which trap heat in our atmosphere, causing temperatures to rise. 2024 saw the fastest rise in atmospheric CO_2 concentrations since records began in 1958, measured at the Mauna Loa observatory in Hawaii. In 2024, CO_2 concentrations reached an average of 424.61 parts per million (ppm), the highest level in at least 800,000 years and rising faster than ever.

The Met Office estimated that the global average temperature for 2024 was around 1.53 °C above the 1850–1900 average. However, this doesn't mean we've failed to keep to the 1.5 °C limit. 2024 was just one year above average and the Paris Agreement specifies a long-term average increase of 1.5 °C. Currently, scientists estimate the global long-term average increase is around 1.3 °C. However, it shows how close we're getting to the limit and highlights, more than ever, that dramatic reductions in emissions are needed.

A brief history of climate change
Since the Industrial Revolution, our world has been warming and our climate changing. The burning of fossil fuels, such as coal and petroleum, releases greenhouse gases and these

INTRODUCTION

create a warming effect in the atmosphere. These gases, including water vapour, carbon dioxide and methane, are naturally present in our atmosphere and are essential to life on Earth. When energy from the Sun reaches the Earth's surface, much of it is reflected as infrared radiation. However, greenhouse gases trap this energy, raising the temperature of the atmosphere; this is called the greenhouse effect. Without them, the Earth's surface temperature would be about −18 °C. The problem is, the more greenhouse gases are present in our atmosphere, the more heat is trapped, so you can have too much of a good thing.

Through the 1900s, concern began to grow about the increasing levels of greenhouse gases and the effect they were having on our climate. Scientists learnt that the amount of carbon dioxide in our atmosphere was directly related to an increase in temperature. Using bubbles of air found in ice cores, they discovered carbon dioxide levels had never been higher in the past 800,000 years. In 2021, the Intergovernmental Panel on Climate Change proved unequivocally that the rise in greenhouse gases was caused by human influence.

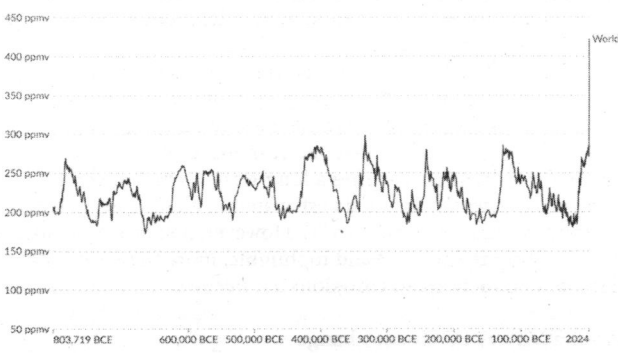

Carbon dioxide concentrations over the past 800,000 years as recorded in ice cores and atmospheric sampling. Data from NOAA global monitoring laboratory, 2025.

In the early 2000s a group of climate scientists published a paper proving a link between climate change and severe weather. It showed that climate change had nearly doubled the chance of the record-breaking heatwave of summer 2003, which resulted in thousands of deaths across Europe. The process of linking a weather event to climate change is known as an attribution study, and this was the first of its kind to be peer-reviewed and published in a scientific journal. Fast forward more than 20 years and attribution studies are much more commonplace, with teams of scientists across the world investigating how climate change impacts severe weather events.

We know that climate change is causing extreme weather such as heatwaves, droughts, storms and flooding to become more frequent and more intense all over the world. Here in the UK, our summers are getting hotter and drier while our winters are becoming warmer and wetter. In fact, a study by the World Weather Attribution group revealed that rainfall during the 2023/24 storm season was 20 per cent heavier because of human-induced climate change. Meanwhile, the record-breaking heatwave of summer 2022, in which the UK measured its first temperature above 40 °C, was at least ten times more likely because of climate change. The Met Office define a heatwave as three or more consecutive days in which a certain temperature threshold is reached. This threshold changes depending on location, and a map of these is shown on the next page. Researchers from the Met Office Hadley Centre warn that if greenhouse gas emissions remain high, 40 °C days could be as frequent as every three or four summers by the end of the century. However, if emissions are reduced, that chance becomes dramatically lower. There is still time to avoid some of these changes. It's easy to feel overwhelmed, but taking steps to lower our emissions can be exciting and empowering. It can be something as simple as digging a garden pond to improve biodiversity or taking public transport instead of driving. Making better choices about the way we eat, shop and travel can have a huge impact. Although there is plenty we can do as individuals, the best way to make a difference is by working together. Spread the word and inspire friends and family to reduce their

INTRODUCTION

emissions. Pressure governments and large businesses to make meaningful changes. Focus on what you can do, instead of what you can't.

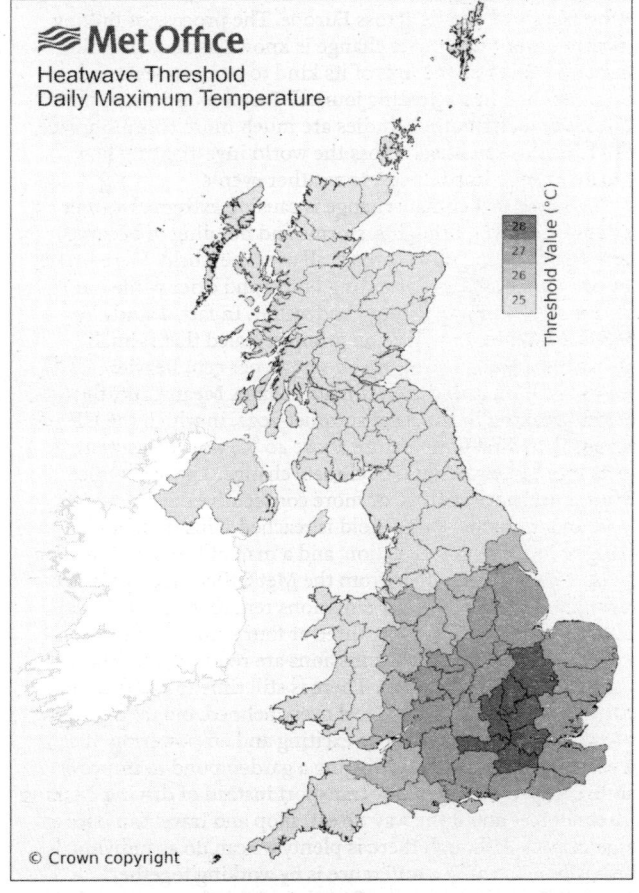

A map of Met Office UK Heatwave Thresholds.

Introduction

In the heart of meteorological winter, January begins. Sparkling frosts cling to plants and windscreens, ice glazes the surfaces of ponds and shades of pale blue, grey and lilac dance across the skies in spectacular cloudscapes. Winter skies are often the most dramatic of the year, lending themselves to stunning photographs and paintings. The secret lies in a handful of ingredients to create these beautiful scenes, the first being the low angle of the Sun during winter. Light from the Sun – which is made up of lots of different colours, each at different wavelengths – has to travel further through the atmosphere to reach us during winter. Blue light, which can't travel as far, is 'scattered', while reds and oranges, which have longer wavelengths, reach the observer. This process, known as Rayleigh scattering, is what gives sunsets such stunning hues. It's also down to the type of cloud; mid- and high-level clouds tend to make the most interesting formations. The final ingredient is cold air, which holds less moisture than warm air so is less humid. It's also generally 'cleaner', as air from the Arctic contains less pollution. The clearer air makes the colours appear more vibrant to us.

On average, January is the coldest month of the year, not only contributing to fantastic cloudscapes, but bringing a plethora of winter conditions, such as snow, ice and freezing rain. Blasts of cold air from the north bring frigid temperatures and snow showers piling in from the North Sea, which can lead to significant snow accumulations, especially inland from north-facing coastlines. Alternatively, if the UK is already in the grip of cold air, low-pressure systems moving in from the Atlantic can also lead to considerable snow along the edge of the leading front. Where it meets the cold air, rain can turn readily to snow, but often turns back to sleet, rain or freezing rain as the system moves through, especially at lower altitudes. Historically, cold spells have persisted for much longer than in recent winters, which have been milder and wetter. In fact, January of 1982 was part of the same cold winter in which the UK recorded its lowest temperature on record, −27.2 °C, on 10 January in Braemar, Aberdeenshire, although it had previously reached this low once before. It wasn't just in

JANUARY

Scotland; temperatures plummeted across the whole of the UK. England also recorded its lowest temperature since records began, -26.1 °C, on the same day in Shropshire. The Met Office daily weather summary described it as 'bitterly cold with a keen easterly wind'. The days leading up to these glacial temperatures saw snowstorms that raged for more than 40 hours in Wales and parts of southwestern Britain. They began late on 7 January, as an occluded front began to push in from the southwest. Snow continued to spread northeastwards into England and Wales overnight, and persisted all day through the 8th and much of the 9th. Snowdrifts up to six metres were reported and many places were left at a complete standstill as rural towns and villages were cut off.

Weather word: Feefle (Scottish)
To swirl, as of snow around a corner.

Weather Extremes in January

Country	Temp.	Location	Date
Maximum temperature			
England	17.6 °C	Eynsford (Kent)	27 January 2003
Wales	18.3 °C	Aber (Gwynedd)	10 January 1971 27 January 1958
Scotland	19.9 °C	Achfary (Sutherland)	26 January 2024
Northern Ireland	16.4 °C	Knocharevan (County Fermanagh)	28 January 2024
Minimum temperature			
England	−26.1 °C	Newport (Shropshire)	10 January 1982
Wales	−23.3 °C	Rhayader (Powys)	21 January 1940
Scotland	−27.2 °C	Braemar (Aberdeenshire)	10 January 1982
Northern Ireland	−17.5 °C	Magherally (County Down)	1 January 1979

Country	Pressure	Location	Date
Maximum pressure			
Scotland	1053.6 hPa	Aberdeen Observatory	31 January 1902
Minimum pressure			
Scotland	925.6 hPa	Ochtertyre (Perthshire)	26 January 1884

JANUARY

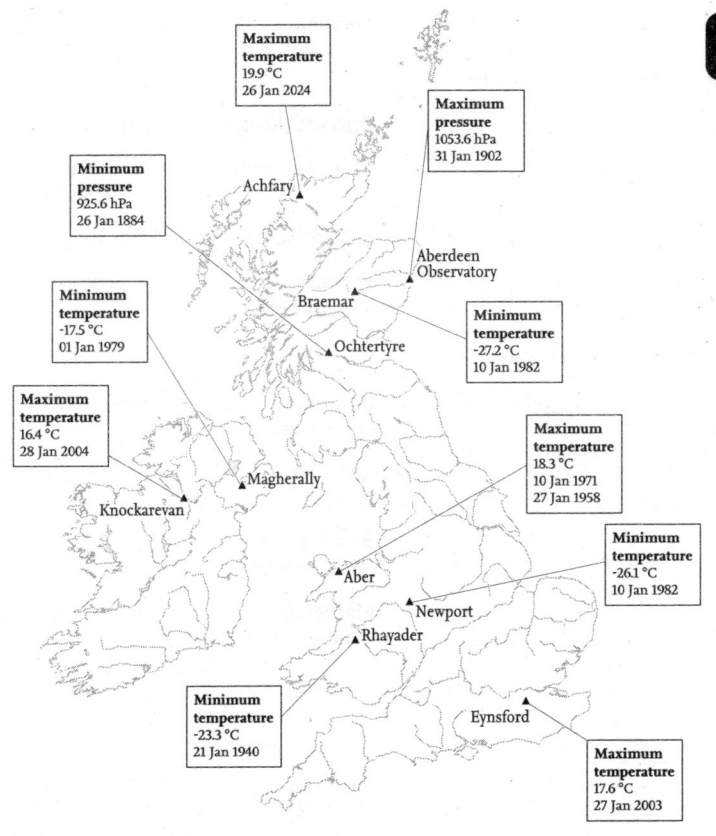

The Weather in January 2024

Observation	Location	Date
Max. temperature 19.9 °C	Achfary (Sutherland)	28 January
Min. temperature −14.0 °C	Dalwhinnie No. 2 (Inverness-shire)	17 January
Most rainfall 127.6 mm	Wet Sleddale Reservoir (Cumbria)	21 January
Most sunshine 8.1 hrs	Oxford (Oxfordshire) Manston (Kent)	18 January 19 January
Highest gust 99 mph (159 kph/86 kt)	Brizlee Wood (Northumberland)	21 January
Greatest snow depth 37 cm	Altnaharra No. 2 (Sutherland)	18 January

January 2024 was a variable month with a real mix of winter weather. It began unsettled, with Storm Henk arriving on the 2nd leading to widespread flooding, especially around the River Trent. By the 5th, most stations across central and southeastern England had recorded two-thirds or more of January's average rainfall already. About a week into the month, high pressure began to build. The resultant set-up saw cold easterly winds and even a rare covering of snow in the Channel Islands.

On the 13th, a cold front swept southwards, bringing a blast of cold Arctic air. Brisk northerly winds brought frequent snow showers to the north, leading to drifts over a metre deep on some roads in the Highlands. Meanwhile, a frontal system moving eastwards on the 16th brought spells of snow to northern England and Scotland. Otherwise, southern England stayed relatively dry, settled and chilly. The cold spell ended with the return of westerly winds around the 19th/20th, followed by the second named storm of January, Storm Isha, arriving on the 21st. Amber and red wind warnings were issued by the Met Office for strong winds. Power cuts affected more than 50,000 homes in northern England and Northern Ireland, with disruption on the rail networks in Scotland and northern England. Hot on the heels of Storm Isha, Storm Jocelyn arrived two days later. Jocelyn was not as intense as Isha but it still led to further disruption across the north of the UK.

Conditions remained mild and unsettled through to the end of January. A strong southwesterly flow on the 28th led to a new daily maximum temperature record in Scotland, thanks to the Foehn effect (see page 25), with 19.9 °C recorded at Achfary in Sutherland. Overall, temperatures in January were close to average with an anomaly of −0.1 °C, with a mild start and end balancing out the mid-month cold spell. Rainfall was also near average at around 97 per cent. Meanwhile, thanks to bright skies during the mid-month cold spell, sunshine hours were above average, at 128 per cent. Globally, 2024 started with the warmest January on record.

The averages are:

Maximum temperature	6.66 °C
Minimum temperature	1.21 °C
Rainfall	121.48 mm
Sunshine	47.45 hrs
Air frost days	11.32

Historic Events

6 January 1998 – A violent tornado, initially developing as a waterspout before moving inland, struck the coastal town of Selsey, West Sussex, just before midnight. Accompanied by golf-ball-sized hail, it ripped off roofs, smashed windows and tore down trees. With a swathe of destruction up to 900 metres wide in places, it was the widest tornado on record.

7 January 1928 – A combination of post-Christmas snowmelt, heavy rain, high spring tides and a storm surge in the North Sea led to catastrophic flooding in London. The River Thames burst its banks, sending a deluge of floodwater into riverside properties, including the Houses of Parliament, the Tate Gallery and the Tower of London. Fourteen people lost their lives while thousands were made homeless.

11–14 January 1987 – Cold easterly winds racing across the relatively warm North Sea picked up plenty of moisture, which then fell as heavy snow over southeastern England from the 11th. More than 50 cm of snow lay in some places, bringing serious disruption as villages were cut off, trains cancelled and schools closed.

23–24 January 1996 – As night fell on the 23rd, freezing rain glazed southern Britain in a thick layer of ice. High pressure and easterly winds in the preceding days had led to very cold air building across the country, but as low pressure moved northwards, rain turned to sleet, snow and freezing rain overnight. Roads became treacherous and power lines were weighed down with ice, causing widespread disruption as dawn broke on the 24th.

JANUARY

31 January 1902 – The UK's highest pressure on record, 1053.6 hPa, was reached in Aberdeen as an intense anticyclone sat over Scotland and southern Norway. At the time, pressure was measured using mercury barometers, with a reading of 31.11 inches recorded at 10 p.m. in Aberdeen.

In this month...

3 January – Wolf Moon

3 January – Perihelion – The Earth is at the closest point to the Sun in its orbit.

18 January – New Moon

Look out for: Foehn effect
When air rises over high ground, it cools and condenses, often bringing some precipitation in the form of rain or snow. As it flows back down the leeward side of the hill or mountain, it warms at a greater rate than it cooled on its ascent, because it is drier. This leads to higher temperatures than on the windward side of mountains. The Foehn effect contributed to January's highest recorded temperature, 19.9 °C at Achfary in Scotland.

Observing the Weather – Frozen Bubbles

You will need:
- A jug
- A paper or reusable silicone straw
- 50 ml warm water
- 2 tsp golden syrup
- 2 tsp washing-up liquid
- 1.5 tsp sugar

In the depths of winter when temperatures plummet into the low minus figures, a rare opportunity to watch ice formation in action arises, in the form of freezing bubbles. The colder the better; temperatures below −8 °C are ideal for this experiment. The best time to try this is after a cold night, first thing in the morning just after the Sun has risen.

First, you need to mix your bubble solution. Add warm water to a jug, followed by golden syrup – which thickens the mixture – then washing-up liquid for the all-important bubble formation, and sugar to help crystals to form. Give it a good mix, then place it in the freezer for 5 minutes while you get ready to go outside.

Once outside, dip your straw into the bubble solution then blow bubbles onto a surface – if you've got lying snow on a bench or windowsill, this is perfect. Blowing the bubbles can be tricky and it may require several attempts to practise your technique so the bubbles don't pop. Once you've cracked it and you've blown a bubble, simply step back and watch in wonder as the bubble freezes before your eyes!

A bubble freezing in the Cairngorms National Park on the morning of 11 January 2025.

JANUARY

Sun and Moon Times in January 2026

Location	Date	Sunrise	Sunset	Moonrise	Moonset
Belfast					
	01 Jan (Thu)	08:46	16:08	13:23	07:18
	11 Jan (Sun)	08:40	16:22	01:48	11:15
	21 Jan (Wed)	08:30	16:40	09:43	20:01
	31 Jan (Sat)	08:14	17:00	14:47	08:01
Cardiff					
	01 Jan (Thu)	08:18	16:14	13:38	06:41
	11 Jan (Sun)	08:14	16:27	01:27	11:14
	21 Jan (Wed)	08:05	16:42	09:25	19:56
	31 Jan (Sat)	07:52	17:00	14:56	07:28
Edinburgh					
	01 Jan (Thu)	08:43	15:49	12:59	07:20
	11 Jan (Sun)	08:37	16:04	01:41	10:59
	21 Jan (Wed)	08:25	16:22	09:36	19:47
	31 Jan (Sat)	08:09	16:43	14:24	08:01
London					
	01 Jan (Thu)	08:06	16:02	13:25	06:28
	11 Jan (Sun)	08:02	02:14	01:14	11:02
	21 Jan (Wed)	07:53	16:30	09:12	19:43
	31 Jan (Sat)	07:40	16:48	14:43	07:15

*all times in GMT

Twilight Times in January 2026

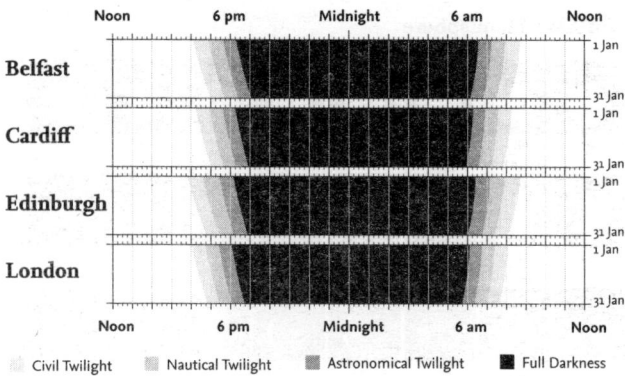

The darkness begins to relinquish its hold as January progresses and days slowly start to lengthen. By the end of the month, the UK will have gained more than an hour of daylight compared to the beginning of January; the change most pronounced in the north where the length of the day increases by more than 90 minutes. But lighter does not always mean warmer. As the old proverb says: 'the cold grows stronger as the days grow longer', meaning the coldest part of the year is yet to come.

Moon Phases in January 2026

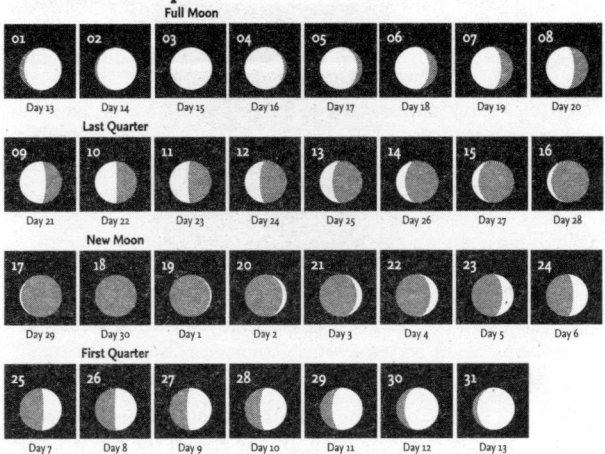

The Wolf Moon
Across Europe, the Full Moon in January is known as the 'Wolf Moon', named after the howls of wolves in midwinter. Despite popular belief, they don't howl at the Moon, rather as a means of communicating with other pack members while warding off rivals. Although wolves were hunted to extinction in Britain at some point in the eighteenth century, their presence here and in Ireland is still remembered by the naming of January's Full Moon.

Other names for January's Full Moon include the Celtic 'Stay Home Moon' and 'Quiet Moon', or the Anglo-Saxon 'Moon after Yule', which refers to the winter solstice festival, falling around 21 December.

WEATHER ALMANAC 2026

Sudden Stratospheric Warmings

During the winter months, meteorologists always keep one eye on the stratospheric polar vortex, an area of very fast winds circling anticlockwise about 50 km above the North Pole. It forms during the autumn, as air above the Arctic begins to cool rapidly, thanks to the descending darkness of the polar night. It's driven by the large difference in temperature between the pole and the mid-latitudes, with the vortex enclosing a pool of very cold air. Although it often persists undisturbed all winter before dissipating in spring, in some years it is disrupted by strong atmospheric waves and can have a knock-on effect on our weather. Some years, the winds weaken or even reverse altogether, triggering a process called a sudden stratospheric warming (SSW) – a jump in temperatures in the stratosphere, which rise by about 50 °C within a few days.

This dramatic phenomenon was first discovered in January 1952 by the German meteorologist, Richard Scherhag. Born in 1907, he had a love of the weather from a young age, installing a climate station in his parents' garden as a teenager. His

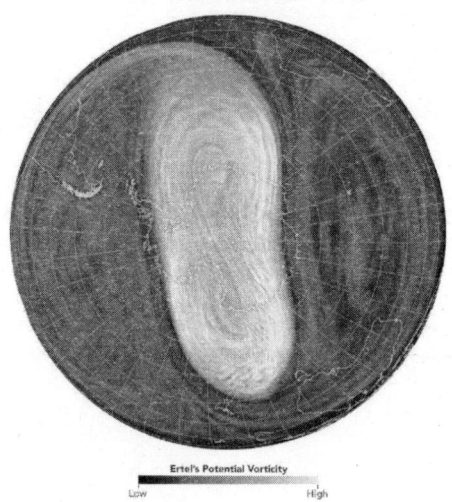

The stratospheric polar vortex.

JANUARY

Richard Scherhag releasing a weather balloon.

speciality was upper-air meteorology, and he was the first to take continued measurements in the upper stratosphere using a radiosonde – a battery-powered instrument attached to a weather balloon that measures a variety of meteorological parameters and sends them back via radio. During his experiments, he became the first to observe an SSW, or as he called it, an 'explosion-like warming of the stratosphere'.

Once the polar vortex is disrupted and an SSW has taken place, it can slowly begin to influence the jet stream, thereby having an impact on our weather. Losing its strength, the jet stream develops a wavy shape, with deep troughs and steep ridges. Becoming almost stationary, it may reduce the frequency of lows coming in from the Atlantic and instead bring much colder air in from the east.

SSWs occur every couple of years but do not always have the same effect. However, a recent event that is sure to have stuck in everyone's minds is the 'Beast from the East' – a period of very cold and snowy weather at the end of February/early March in 2018. It was preceded by an SSW in January of the same year, which eventually weakened the jet stream, allowing very cold air from the heart of Siberia to bring sub-zero temperatures and heavy, disruptive snow to the UK.

Introduction

From winds blowing through the bare trees and rain lashing at the windowpanes, to blankets of snowdrops in full bloom, steeped in bright winter sunshine, February's weather can be very mixed. As winter draws to a close, the weather in its final month is often dictated by the position of the jet stream. Strongest in winter, thanks to the temperature difference between the Equator and the North Pole, it is responsible for bringing wet weather in the form of low-pressure systems in from the Atlantic. It also marks the boundary between mild air to the south and cold air to the north. Therefore, when the jet stream is directly over the UK, we tend to see mild and wet weather. The winter of 2014 was one such wet and windy winter. From late January to mid-February that year, six major storms hit at intervals of two or three days, with those arriving in February remembered as being particularly severe. It was the wettest winter on record, with major flooding issues across the UK. The series of deep depressions was associated with an unusually powerful jet stream, driven by a strong temperature contrast across North America.

However, if the jet stream slips southward, the UK can be plunged into cold air and often settled weather on the northern side of the jet. The reverse is also true: should the jet stream be positioned well to the north of the UK, settled weather can also prevail, but it will be milder. Research into how climate change will affect the jet stream is of great importance. However, there is still some debate over the results. Some scientists suggest that the rapidly warming poles may weaken the jet stream as the temperature contrast between the Equator and high latitudes lessens. In this scenario, the jet stream would slow, buckle and become more wavy, leading to a greater chance of blocking highs (see page 57). In winter, this can bring long spells of very cold weather.

To throw another weather type into the mix, mid-February also marks the return of land-based convection. To the weather watcher, this means welcoming convective cumulus clouds back into our skies, and the chance of showers. This is thanks to the increasing warmth from the Sun, heating the air close to the ground and causing it to rise. As the air rises, it condenses, forming the fluffy clouds we know and love and perhaps even triggering the odd rumble of thunder.

In winter, thunderstorms are more likely to develop in squall lines along the edge of a cold front. This process is also known as line convection, which develops thanks to rapidly rising air along a frontal boundary, creating a narrow band of intense rainfall. This effect led to severe thunder and hailstorms across southern England and Wales on 22 February in 1995 as a cold front swept southeastwards. The squall line was described as 'a wall of torrential rain or hail' and was accompanied by violent gusts of wind. There were also reports of tornadoes, although none were confirmed.

Weather word: Dreich (Scottish, fifteenth century)
Dreary, cheerless and bleak weather, sometimes with the implication that the rain is fine, but falling densely and continuously.

Weather Extremes in February

Country	Temp.	Location	Date
Maximum temperature			
England	21.2 °C	Kew Gardens (London)	26 February 2019
Wales	20.8 °C	Porthmadog (Gwynedd)	26 February 2019
Scotland	18.3 °C	Aboyne (Aberdeenshire)	21 February 2019
Northern Ireland	17.8 °C	Bryansford (County Down)	13 February 1998
Minimum temperature			
England	−22.2 °C	Scaleby (Cumbria) Ketton (Leicestershire)	19 February 1892 8 February 1895
Wales	−20.0 °C	Welshpool (Powys)	2 February 1954
Scotland	−27.2 °C	Braemar (Aberdeenshire)	11 February 1895
Northern Ireland	−15.6 °C	Garvagh, Moneydig (County Londonderry)	20 February 1955

Country	Pressure	Location	Date
Maximum pressure			
Scotland	1052.9 hPa	Aberdeen Observatory	1 February 1902
Minimum pressure			
Republic of Ireland	942.3 hPa	Midleton (County Cork)	4 February 1951

FEBRUARY

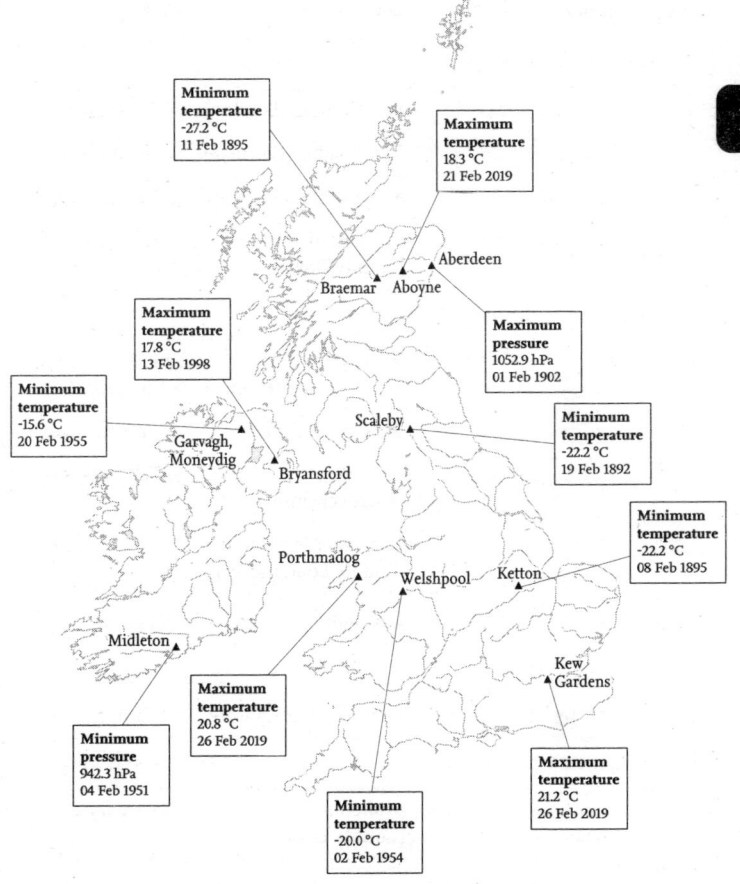

The Weather in February 2024

Observation	Location	Date
Max. temperature 18.1 °C	Pershore College (Hereford & Worcester) and Teddington, Bushy Park (Middlesex)	15 February
Min. temperature −13.8 °C	Altnaharra No. 2 (Sutherland)	8 February
Most rainfall 86.9 mm	White Barrow (Devon)	17 February
Most sunshine 9.6 hrs	Kinross (Kinross-shire)	29 February
Highest gust 72 mph (115 kph/63 kt)	Capel Curig No. 3 (Gwynedd)	21 February
Greatest snow depth 15 cm	Kirkwall (Orkney)	7 February

Unsettled weather continued into February, which was a very mild and wet month overall. It began with high pressure centred to the south, bringing a relatively warm westerly flow to the UK. After several days, things started to change as a cold front spread southwards, introducing chilly conditions and spells of snow in the north. Meanwhile, rain fell in the south and from the 6th to the 8th a trailing frontal system brought flooding to southern areas, before rain became more widespread into the 9th. By mid-month, some weather stations in the south had already recorded a month's worth of rainfall. As well as being wet, southern areas generally stayed on the milder side throughout February. Within a week, temperatures in the north had become closer to average and most places found themselves in a pattern of wet and mild weather.

Further flooding occurred across England and Wales in the last week of February, thanks to heavy rain on the 22nd and a low-pressure system on the 25th bringing rain to the far south of England. Overall, it was the fourth-wettest February on record for England, which received nearly twice its average rainfall. Some stations in the south received more than three times their average rainfall, and for southern England and East Anglia, it was the wettest February on record. Thanks to a predominantly westerly airflow bringing in mild and moist air from the Atlantic, it was also 2.2 °C warmer than average across the UK, the second warmest after 1998 and the warmest on record for England and Wales. With cloudy skies and frequent rain, February was duller too, with only 79 per cent of average sunshine.

February was fairly representative for winter 2023/24 as a whole, coming in fifth warmest and eighth wettest on record. Mild, wet winters, such as this one, illustrate clearly the effect of climate change here in the UK, with warm air able to hold more moisture, leading to more rain.

The averages are:

Maximum temperature	7.16 °C
Minimum temperature	1.13 °C
Rainfall	96.15 mm
Sunshine	71.86 hrs
Air frost days	10.69

WEATHER ALMANAC 2026

Historic Events

5 February 1683 – A rapid thaw of snow and ice at the end of a long cold spell led to widespread flooding in the Trent valley. Melting snow and river ice formed fast-moving ice floes which destroyed bridges in Nottingham and Newark-on-Trent.

11 February 1991 – Infamous newspaper headline 'The Wrong Type of Snow' was born during a spell of severe wintry weather which caused widespread disruption on the train lines. Cold easterly winds brought temperatures well below freezing and the resultant snowfall was dry, powdery and easily blown around, and into the electrical systems of trains. Attempting to explain the issue, British Rail coined the phrase during a radio interview and was subsequently ridiculed.

13 February 1997 – Stormy weather approaching southwestern England hit the Tokio Express container ship, with a rogue wave washing 62 containers overboard. One of these was full of nearly 5 million pieces of Lego, destined to be made into toy kits. In a bizarre coincidence, the pieces included mainly sea-themed items, such as octopuses, scuba gear, flippers and ship rigging. They've been washing up on beaches in Cornwall and Devon ever since and have even been found as far away as France, Belgium and the Netherlands.

16 February 1962 – A deep depression tracking to the north of Scotland brought exceptionally strong winds to much of central and northern Britain and was dubbed the 'Great Sheffield Gale' for the destruction it caused in the English city. About two-thirds of houses there suffered damage as winds gusted up to 97 mph in the city.

FEBRUARY

26 February 1903 – One of the most damaging extratropical cyclones to ever affect the UK, Storm Ulysses brought widespread gusts in excess of 100 mph. It left a trail of destruction, including shipwrecks, fallen trees and derailed trains. It went down in history, even receiving a mention in James Joyce's 1922 novel, *Ulysses*, for which the storm was subsequently named.

In this month...

1 February – Snow Moon

5 February – National Weatherperson's Day

11 February – International Day of Women and Girls in Science

17 February – New Moon

Look out for: Fogbows
This phenomenon, sometimes known as a white rainbow, forms in a similar way to its multicoloured sibling but as the name suggests, in fog rather than rain. It develops when sunlight interacts with the tiny droplets of water present in fog. These are much smaller than raindrops, so light is diffracted as well as refracted (the process which happens in a traditional rainbow), giving the fogbow its distinctive white colour.

Observing the Weather – Snow Depth

Most official weather stations measure snow depth using ultrasonic sensors, which are automated and very accurate. Before these were invented, there was a much simpler way to measure snow and it's easy to try at home, provided you've got some lying snow! You will also need a ruler (one that starts with zero at the very end), tape measure or measuring stick.

One of the most important parts of measuring snow depth is your choice of location. Pick somewhere away from trees, tall buildings or snow drifts, ideally on a flat, level surface, like a patio or decking. Hold the ruler vertically and push it into the snow until you feel it has reached the ground. Read the measurement in centimetres, to the nearest tenth. If you want to be even more accurate, take two more measurements at different locations and average the three readings by adding them together and dividing by 3.

Snow depth measurement taken in Sheerness, Kent, on 27 February 2018 during the 'Beast from the East'.

Sun and Moon Times in February 2026

Location	Date	Sunrise	Sunset	Moonrise	Moonset
Belfast					
	01 Feb (Sun)	08:12	17:02	16:23	08:25
	11 Feb (Wed)	07:53	17:22	04:37	10:32
	21 Feb (Sat)	07:31	17:43	08:25	23:31
	28 Feb (Sat)	07:15	17:57	13:54	06:29
Cardiff					
	01 Feb (Sun)	07:50	17:02	16:27	07:58
	11 Feb (Wed)	07:34	17:20	04:00	10:47
	21 Feb (Sat)	07:14	17:38	08:21	23:09
	28 Feb (Sat)	07:00	17:51	14:01	05:59
Edinburgh					
	01 Feb (Sun)	08:07	16:45	16:03	08:23
	11 Feb (Wed)	07:46	17:07	04:39	10:07
	21 Feb (Sat)	07:23	17:29	08:11	23:24
	28 Feb (Sat)	07:06	17:44	13:34	06:28
London					
	01 Feb (Sun)	07:38	16:49	16:13	07:45
	11 Feb (Wed)	07:21	17:08	03:47	10:34
	21 Feb (Sat)	07:02	17:26	08:09	22:56
	28 Feb (Sat)	06:47	17:38	13:48	05:47

*all times in GMT

Twilight Times in February 2026

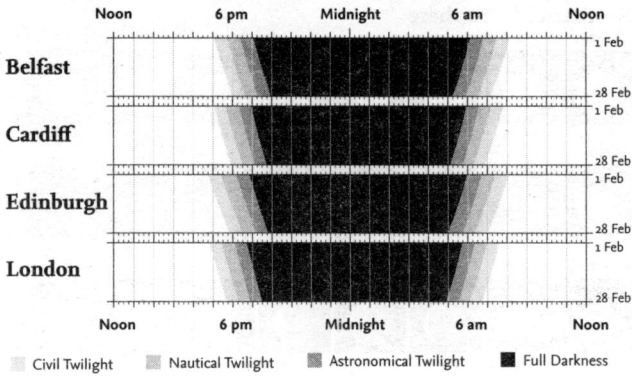

Daylight hours continue to lengthen in earnest through February. For much of Scotland, an additional two hours is gained by the end of the month as darkness loosens its grip. Meanwhile, further south daylight also increases, although the change is slightly less marked, with some places only gaining an extra hour and a half.

Moon Phases in February 2026

Northern Hemisphere

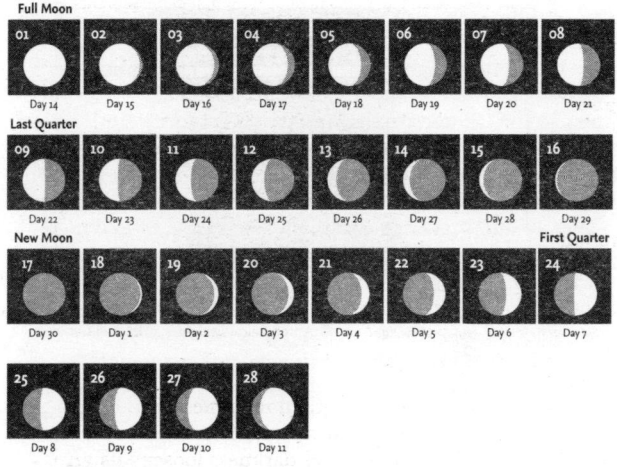

The Snow Moon
February's Full Moon is known as the 'Snow Moon', for this is the month in which lying snow is most likely, based on historical averages. Alternative Celtic and Old English names for the February Moon are the 'Storm Moon' and 'Ice Moon', both also referring to the typical wintry weather of the month. Meanwhile, another Native American name is 'Hunger Moon', after the scarce food in midwinter.

Winter 1947

Ask anyone over the age of 85 about their most memorable winter and they're sure to mention that of 1947. Snow drifts buried cars and towered along the sides of roads, power lines sagged under the weight of ice and snowfall, and severe frosts gripped the country. It was the snowiest winter of the twentieth century in the UK and particularly tough as World War Two had only ended two years previously, and rationing was still in place. The peak of the exceptionally cold and snowy weather was February, the second coldest since records began.

After an unremarkable start to the winter and several mild spells through January, including one in which temperatures rose up to around 14 °C, the cold weather began to take shape around the 20th as an area of high pressure drifted up from France, to become centred over the North Sea. The resultant easterly winds introduced very cold air from continental Europe and on the evening of 22 January snow began to fall in the east. From then until 17 March, snow fell somewhere in the UK every day and with bitterly cold conditions throughout, it soon built up.

Through February, temperatures rarely rose more than a degree or two above freezing and snow lay on the ground for the whole month over much of the country. It was also persistently dull, with no sunshine recorded at Kew Observatory from the 2nd to the 22nd. When skies did clear, a minimum of −21 °C was recorded at Woburn in Bedfordshire. Snow was heaviest in the east as air travelling across the North Sea was able to pick up plenty of moisture, which upon reaching the east coast, fell in earnest as snow. Throughout the month, there were several snowfalls of 60 cm or more and drifts of up to five metres blocked roads and railways. This in turn caused problems with the transport of fuel, leading to power shortages. Radio and TV broadcasts were limited or even suspended completely, while newspapers had to be reduced in size. Farmers lost livestock and crops were destroyed. On 13 February, a Halifax bomber crashed on a mission to deliver supplies to stranded communities in the Peak District, killing six crew and two journalists on board.

Heavy snowstorms continued into March, with drifts of up to seven metres reported in the Scottish Highlands. The thaw began around 10/11 March and progressed rapidly across the UK. Widespread flooding ensued as hard, frozen ground was unable to absorb the meltwater. Instead, it poured into rivers, bursting their banks and flooding hundreds of homes. The government drafted in troops to try to control floodwaters in the East Anglian Fens. Floods began to subside towards the end of March, but further rainfall did little to help the rate of decrease.

Towering snow drifts brought travel to a standstill, as shown by this double-decker bus buried in snow.

Introduction

Once-bare trees begin to bud, cloudy skies break to reveal the pleasant warmth of March sunshine and persistent rain eases to something more showery. Meteorological spring is here, a true season of transition. And as March progresses, the weather is changing too. Popular weather lore describes this month as 'in like a lion, out like a lamb', and although in many years the opposite happens or something completely different occurs, this no doubt refers to the idea that March is a transitional month between the winter storms of February and the generally more settled weather of April.

Despite the picture of fair spring weather that comes to mind, March often continues where winter left off, with further Atlantic depressions tracking eastwards across the UK, bringing a continuation of wet and windy weather. Sometimes this can persist for the whole of the month, as it did in the UK's wettest March on record in 1981. Weather archives describe it as a 'very wet, very dull, very mild' month with frequent depressions crossing the country and a mild, moist, southwesterly airflow. Forced to rise and release its rainfall as it met high ground in Wales, the resultant orographic effect gave these areas some of the highest rainfall totals, particularly on the 21st and 22nd, when a report from Eryri (Snowdonia) suggested 129 mm of rain had fallen within the 24 hours from 9 a.m. on the 21st.

Since the Met Office began naming storms in 2015, there have only been three named in March: Storm Jake in 2016, which brought strong winds to the southwest of the UK; Storm Freya in 2019, accompanied by strong winds, coastal flooding and snow across the northern Pennines; and Storm Gareth, also in 2019 and perhaps the most impactful. Gareth brought powerful winds and heavy rain to Ireland and Northern Ireland initially, before spreading across Britain and into Europe.

When the Sun does come out, this is when March feels its most spring-like, as the increasing heat reaching us from the Sun becomes very noticeable at this time of year. The warmest March day on record was on the 29th in 1968, when temperatures reached 25.6 °C at Mepal in Cambridgeshire. This example teaches us to not be lulled into a false sense of security though, as spells of very cold weather are still possible.

Just a few days after the temperature record, a cold front swept southwards bringing bitterly cold Arctic air and snow to many places.

While it can be variable in terms of weather, March can typically be relied upon to have the coldest sea temperatures on average. This is due to the high thermal capacity of water, which means it takes a long time to heat up and cool down, in comparison to air, which has a low thermal capacity. Although the peak of the cold weather tends to be sometime in winter, the sea will continue to cool and won't start its gradual warming process until April. Certainly something to keep in mind if you fancy a dip on one of March's milder days.

Weather word: Fierce mild (Irish-English)
A fine, warm day.

Weather Extremes in March

Country	Temp.	Location	Date
Maximum temperature			
England	25.6 °C	Mepal (Cambridgeshire)	29 March 1968
Wales	23.9 °C	Prestatyn (Denbighshire) Ceinws (Powys)	29 March 1965
Scotland	23.6 °C	Aboyne (Aberdeenshire)	27 March 2012
Northern Ireland	21.8 °C	Armagh (County Armagh)	29 March 1965
Minimum temperature			
England	−21.2 °C	Houghall (County Durham)	4 March 1947
Wales	−21.7 °C	Corwen (Denbighshire)	3 March 1965
Scotland	−22.8 °C	Logie Coldstone (Aberdeenshire)	14 March 1958
Northern Ireland	−14.8 °C	Katesbridge (County Down)	2 March 2001

Country	Pressure	Location	Date
Maximum pressure			
Republic of Ireland	1051.2 hPa	Malin Head (County Donegal) South Uist (Outer Hebrides)	29 March 2020
Minimum pressure			
Scotland	946.2 hPa	Wick (Caithness)	9 March 1876

MARCH

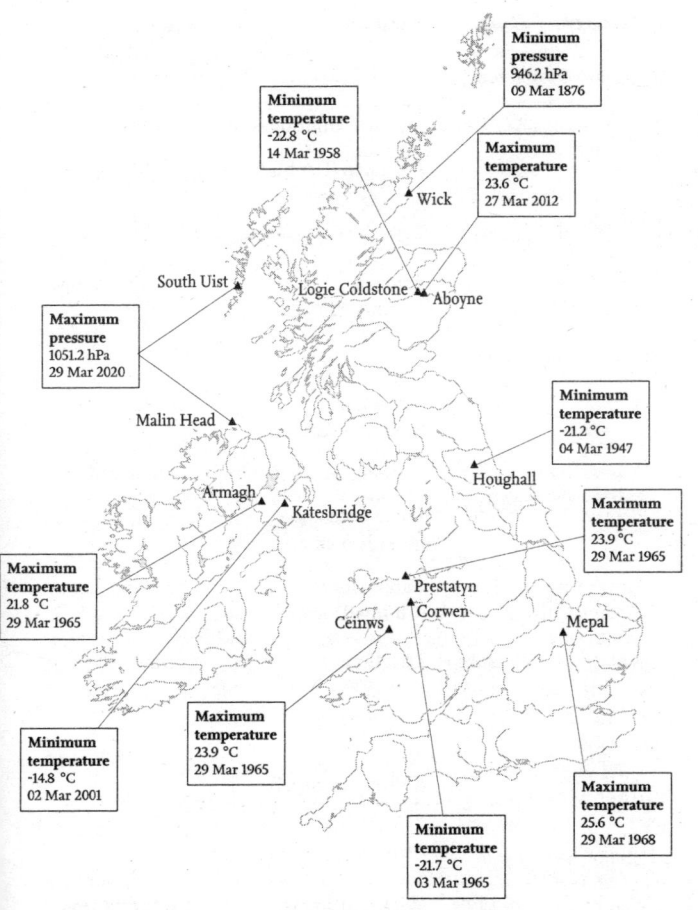

The Weather in March 2024

Observation	Location	Date
Max. temperature 18.8 °C	Charlwood (Surrey)	20 March
Min. temperature −6.9 °C	Altnaharra No. 2 (Sutherland)	26 March
Most rainfall 94.4 mm	Capel Curig No. 3 (Gwynedd)	13 March
Most sunshine 11.5 hrs	Kirkwall (Orkney) Bewcastle (Cumbria)	30 March 31 March
Highest gust 81 mph (130 kph /70 kt)	Needles Old Battery (Isle of Wight) Needles Old Battery (Isle of Wight)	1 March 28 March
Greatest snow depth 12 cm	Lough Navar Forest (Fermanagh)	1 March

Overall a mild, wet and dull month, March nevertheless had some notable cold spells at the beginning and end. Snow swept across Ireland, Wales, the Peak District and the south Pennines on the 1st, followed by heavy showers across southern England and Wales, occasionally thundery with a mixture of hail, sleet and snow over high ground.

The weather remained chilly over the next few days, with further wintry precipitation as a low-pressure system cleared eastwards. Traffic was disrupted in places, with snow ploughs called into action to clear the M4 on the 2nd.

About a week into March, a broad area of high pressure built over Scandinavia, extending across the UK and bringing a brief spell of settled weather. Brisk east-to-southeasterly winds brought a marked wind-chill to many areas. From the 13th, Atlantic westerlies returned, bringing in a series of low-pressure systems. It was mild, wet and windy with temperatures reaching the mid-to-high teens at times. Southern areas were particularly wet and by mid-March, some counties in the southeast had already received a month's worth of rainfall. Following a wet winter, which meant the ground was already saturated with water, groundwater and surface-water flooding was common, especially in Suffolk, where it led to travel disruption. A deep low arriving at the end of the month brought a wet and windy start to Easter weekend. As it passed over the Isles of Scilly, St Mary's weather station recorded its lowest March pressure on record, 963.4 hPa. It also brought spells of sleet and snow to parts of southwestern England overnight on the 27th/28th, leaving an unsettled weekend to follow, with widespread showers and strong winds.

Overall, temperatures were 1.0 °C above average across the UK in March. With frequent wet weather, the UK recorded 127 per cent of its average rainfall, a lot of this falling across southern areas. It was also a dull month, with only 87 per cent of the average sunshine for March. The winter half-year from October 2023 to March 2024 was the wettest six months on record for England and Wales.

The averages are:

Maximum temperature	9.22 °C
Minimum temperature	2.19 °C
Rainfall	85.07 mm
Sunshine	109.18 hrs
Air frost days	7.84

Historic Events

4 March 1970 – Widespread snow fell across southern Britain, with 35 cm recorded in Bedford. In Kent, 700 miners were trapped 915 metres underground for 14 hours as the blizzards cut off power.

9–13 March 1891 – Five days of blizzards hit southern Britain, striking the southwest particularly hard. It was dubbed 'The Great Blizzard of 1891' and saw more than 200 deaths, many of these at sea. Meanwhile, around 6,000 animals died and deep drifts buried stranded trains.

22 March 2013 – An active frontal system spreading up from the southwest met cold easterly winds, and rain turned readily to snow across much of northern Britain and Ireland. Snow reaching 30–40 cm deep fell widely, with drifts over two metres. Meanwhile, southwestern areas saw widespread flooding.

24 March 1878 – A ferocious squall and snowstorm sank HMS *Eurydice* off the coast of the Isle of Wight. There were only two survivors amidst a crew of more than 350. Developing over northern England and spreading southwards throughout the day, it hit the south coast later that evening.

29 March 1958 – The first atmospheric carbon dioxide measurement in a 68-year series, known as the Keeling Curve, was taken. Established by Charles Keeling, the site at Hawaii's Mauna Loa observatory has been measuring CO_2 concentrations almost continuously since then. The initial measurement put CO_2 at 313 ppm, but it has now exceeded 425 ppm.

MARCH

In this month...

1 March – Meteorological spring begins

3 March – Worm Moon

19 March – New Moon

20 March – Astronomical spring begins

20 March – Spring equinox

23 March – World Meteorological Day

28 March – Earth Hour at 8:30 p.m.

29 March – British Summer Time begins

Look out for: Blocking highs
As spring progresses and the relentless succession of Atlantic depressions begins to ease, anticyclones become more likely and 'blocking highs' may develop. These are large, stationary areas of high pressure, which effectively 'block' the eastward progression of low-pressure systems. There are two types of block: omega, after the Greek letter which the blocking pattern resembles, and diffluent, with high pressure to the north and low pressure to the south.

Observing the Weather – Cloud Cover

Whether there's a thick layer of stratus or just a few fluffy cumulus clouds, measuring cloud cover is an important part of taking weather observations. Meteorologists do this using oktas, which are a representation of the fraction of the sky covered by cloud of any type or height, in relation to an observer. Each okta indicates clouds covering an eighth of the sky and follows the guidelines below, as per the World Meteorological Organization:

- 0 oktas – complete absence of cloud, clear skies and fine weather
- 1 okta – fine with one eighth or less of cloud, but not zero
- 2 oktas – fine with two eighths of cloud cover
- 3 oktas – partly cloudy with three eighths of cloud cover
- 4 oktas – partly cloudy with four eighths of cloud cover
- 5 oktas – partly cloudy with five eighths of cloud cover
- 6 oktas – cloudy with six eighths of cloud cover
- 7 oktas – cloudy with seven eighths or more, but not full cloud cover
- 8 oktas – overcast with full cloud cover
- 9 oktas – sky obscured by fog, heavy snowfall or a similar meteorological phenomenon

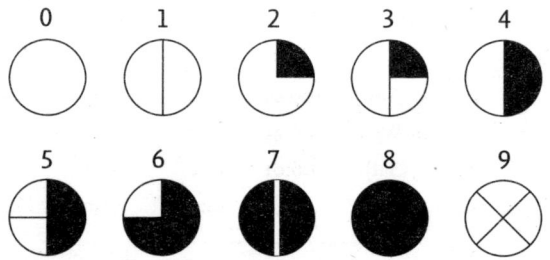

Cloud cover in oktas.

MARCH

Sun and Moon Times in March 2026

Location	Date	Sunrise	Sunset	Moonrise	Moonset
Belfast					
	01 Mar (Sun)	07:13	17:59	15:27	06:47
	11 Mar (Wed)	06:48	18:19	03:28	09:07
	21 Mar (Sat)	06:23	18:39	06:45	22:44
	31 Mar (Tue)	06:58	19:58	18:23	06:25
Cardiff					
	01 Mar (Sun)	06:57	17:52	15:28	06:23
	11 Mar (Wed)	06:35	18:10	02:49	09:23
	21 Mar (Sat)	06:13	18:27	06:44	22:18
	31 Mar (Tue)	06:50	19:44	18:12	06:12
Edinburgh					
	01 Mar (Sun)	07:04	17:46	15:09	06:43
	11 Mar (Wed)	06:38	18:07	03:31	08:41
	21 Mar (Sat)	06:12	18:28	06:29	22:39
	31 Mar (Tue)	16:46	19:48	18:11	06:15
London					
	01 Mar (Sun)	06:45	17:40	15:15	06:10
	11 Mar (Wed)	06:23	17:58	02:37	09:10
	21 Mar (Sat)	06:01	18:15	06:31	22:05
	31 Mar (Tue)	06:38	19:32	17:59	06:00

*times change to BST on 29 March

Twilight Times in March 2026

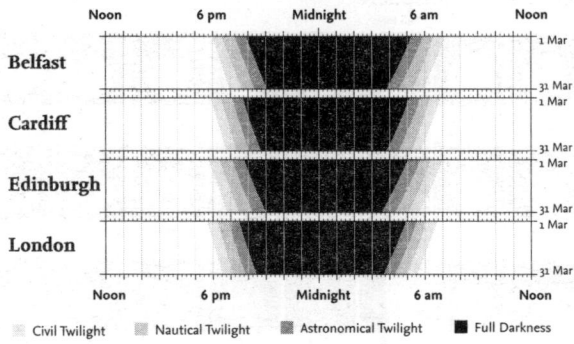

The spring equinox on 20 March marks the moment that the centre of the Sun crosses the Equator from the southern to the northern hemisphere. Derived from Latin, equinox means 'equal night', as day and night are supposedly equal on this day. But thanks to the size of the Sun and the way its light is refracted by our atmosphere, equal day and night actually happens a few days before, on the equilux, which means 'equal light'.

Soon after the equinox the clocks will change, springing forward an hour. From the 29th, the Sun will set an hour later in the evenings.

MARCH

Moon Phases in March 2026

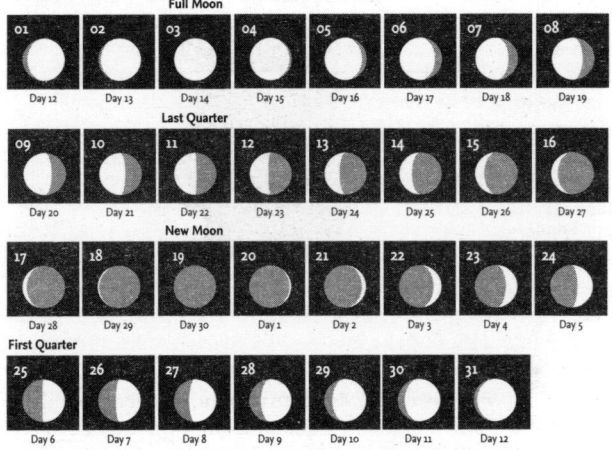

The Worm Moon
As the seasons change and early spring sunshine begins to warm the ground, the first earthworms appear from their winter slumber. March's Full Moon takes its name from this sign of spring and is known as the 'Worm Moon'. It's also known as the 'Lenten Moon', after the lengthening days for which the Christian fast of Lent is also named. Meanwhile, Native American people called this moon the 'Sap Moon', for when maple tree sap runs.

COP1 – Berlin, 1995

Towards the end of the twentieth century, concerns about climate change were growing rapidly. As scientists continued to investigate our warming planet and evidence piled up, theory turned to fact: human activity was causing the rise in temperature. Experts and politicians began calling for something to be done. In the early Nineties, the first international treaty to prevent climate change was signed and the United Nations Framework Convention on Climate Change was established. All member countries agreed to meet every year and the Conference of Parties (COP) was born. On 28 March 1995 the very first conference was held: COP1. Hosted in Berlin, countries discussed the reduction of greenhouse gas emissions and put together a series of commitments, known as the Berlin Mandate. Climate scientists and environmental campaigners criticized the mandate for its lack of firm pledges, thus a group was created to draw up a more specific set of targets and timescales. They returned two years later to COP3, in Kyoto, Japan, and the Kyoto Protocol was proposed. It included legally binding emissions targets and placed the responsibility of reduction on developed countries. The Kyoto Protocol came into force in 2005, after years of trying to persuade members to agree and sign. The protocol was widely hailed as a key moment in the history of climate conferences, although critics continued to question its effectiveness as two of the largest emitters, the US and China, refused to be part of it. Over the following years, many COPs passed with little consequence. In 2007, at COP13, it was agreed that the Kyoto Protocol must be replaced by the time of COP15, but when that came around in Denmark in 2009, countries were unable to reach an agreement. The Kyoto Protocol was rebranded to the much weaker Copenhagen Accord and the conference was described as a failure. The following years were spent trying to 'save' the Kyoto Protocol. Officials argued over the need for a new climate treaty and at COP19 in Warsaw, Poland, 133 countries and hundreds of environmental campaigners walked out over a 'lack of seriousness'. Finally, at COP21 in France, a landmark moment was reached. The Paris

Agreement was unanimously signed in 2015 – a treaty aiming to limit the rise in global temperatures to 1.5 °C by reducing emissions. The next few conferences were spent working through the technicalities of the agreement. Meanwhile, the focus of the most recent COP summits has been the transition away from fossil fuels and 'loss and damage' funding for vulnerable countries. In 2024, global temperatures reached 1.5 °C above pre-industrial levels for the first time. In January 2025, President Trump pulled out of the Paris Agreement for the second time. As our planet continues to warm and extreme weather events unfold around the world, all eyes are on COP30 and COP31 to provide some answers.

Delegates of the first UN climate conference during the opening event on 28 March 1995.

Introduction

Everywhere you look in April, there are signs of new life. Increasing warmth and daylight hours bring magnificent floral displays of blossom, daffodils, bluebells and tulips into bloom. Typically, the weather becomes more settled as large blocking anticyclones develop. These features are more likely to occur in spring, especially April and May, and also bring a shift in prevailing wind direction. Generally, our winds come from the southwest for most of the year. However, during mid-to-late spring, anticyclones developing to the northwest of the UK bring a greater frequency of winds from the northeast. This in turn can create the perfect conditions for a handful of weather phenomena most likely at this time of year, all to do with northeasterly winds. The Helm Wind is a local wind that develops along the North Pennines in northern England. It tends to develop over Cross Fell, the highest peak in the mountain range. When northeasterly winds meet the perpendicularly oriented mountain range, they are forced upwards. Encountering a stable layer of air that acts like a lid, they are then squeezed as they move over. As the air descends the southwestern slope of Cross Fell, pressure is released and the winds speed up, creating the strong, gusty Helm Wind.

When these winds blow, a heavy bank of stationary cloud develops along the Cross Fell range, known as the Helm. It is accompanied by the Helm Bar, another roll of cloud parallel to the Helm three or four miles from the foot of the fell. This is because once the air descends the fell, it rises again, creating a waveform with the cloud at the crest.

The Helm Wind is said to be the only named wind in the UK. However, locals living on the northeast coast of England have affectionately named another cold northeasterly wind that occurs around springtime, 'Custard' winds (see page 67), while those in East Anglia also have their 'Fen Blows': gusty winds that lift freshly cultivated topsoil in spring.

Warmth continues to build through April and although unusual, temperatures into the high 20s Celsius are possible. The highest temperature ever recorded in April was in 1949, when the mercury soared to 29.4 °C at Camden Square in London, although there are some reservations about the reliability of this record. It was the warmest Easter ever recorded too, but it's worth remembering that the Christian celebration can fall on a variety of different dates from late March to late April.

At the other extreme, April is not out of the woods yet in terms of the possibility of cold spells. To quote Eighties pop legend Prince, 'sometimes it snows in April'. Winter can make a dramatic reappearance, as it did in April 1981, with a blizzard remembered for its intensity and exceptional lateness in the season. With cold air already in place across the UK, a complex low-pressure system arriving on the 24th brought heavy snow to much of Scotland, Northern Ireland and northern England. Up to 20 cm fell in places and snow drifted up to one metre in strong northeasterly winds. Over the following days, the snow moved further south bringing 10–20 cm of snow widely to the Midlands, central-southern England and the West Country. With strong winds, the resultant blizzard wreaked havoc on traffic, electricity supplies and the farming community.

Weather word: Custard wind (English)
Cold easterly winds on the northeastern coast of England in spring. Sometimes accompanied by sea fret (see page 83) and thought to have come from the word 'coastward'.

Weather Extremes in April

Country	Temp.	Location	Date
Maximum temperature			
England	29.4 °C	Camden Square (London)	16 April 1949
Wales	26.2 °C	Gogerddan (Ceredigion)	16 April 2003
Scotland	27.2 °C	Inverailort (Inverness-shire)	17 April 2003
Northern Ireland	24.5 °C	Boom Hall (County Londonderry)	26 April 1984
Minimum temperature			
England	−15.0 °C	Newton Rigg (Cumbria)	2 April 1917
Wales	−11.2 °C	Corwen (Denbighshire)	11 April 1978
Scotland	−15.4 °C	Eskdalemuir (Dumfriesshire)	2 April 1917
Northern Ireland	−8.5 °C	Killylane (County Antrim)	10 April 1998

Country	Pressure	Location	Date
Maximum pressure			
Scotland	1044.5 hPa	Eskdalemuir (Dumfriesshire)	11 April 1938
Minimum pressure			
Republic of Ireland	952.9 hPa	Malin Head (County Donegal)	1 April 1948

APRIL

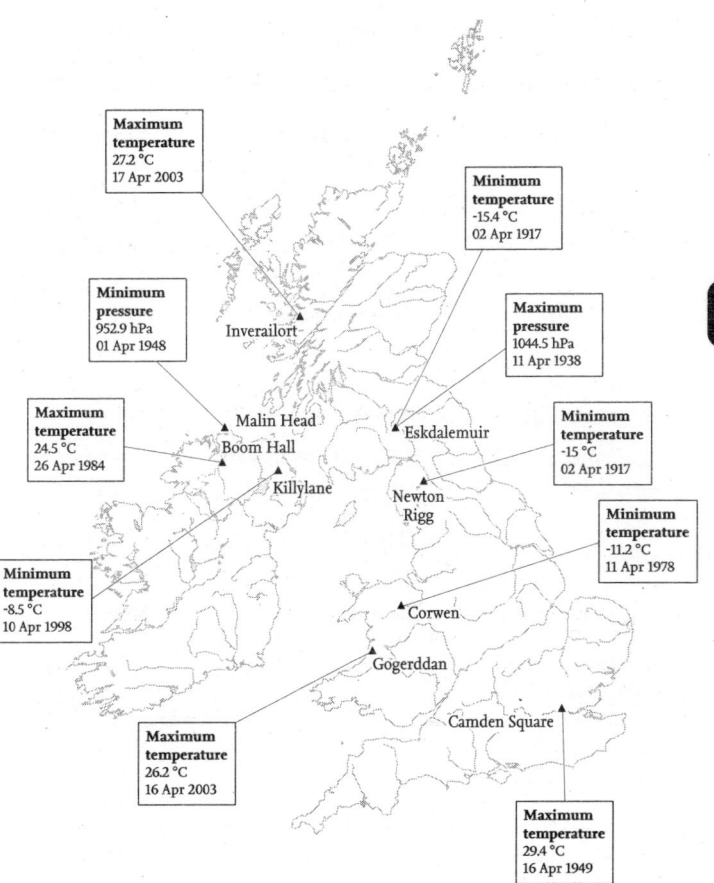

The Weather in April 2024

Observation	Location	Date
Max. temperature 21.8 °C	Writtle (Essex)	13 April
Min. temperature −6.3 °C	Shap (Cumbria)	26 April
Most rainfall 64.8 mm	Mickleden, Middlefell Farm (Cumbria)	29 April
Most sunshine 13.8 hrs	Manston (Kent)	29 April
Highest gust 76 mph (122 kph/66 kt)	Loch Glascarnoch (Ross & Cromarty)	7 April
Greatest snow depth 2 cm	Kinross (Kinross-shire)	5 April

Living up to its reputation, April brought a classic mixture of spring weather, including early warmth, extratropical cyclones and snow showers. Over the first few days low pressure to the south of the UK brought plenty of April showers and occasional sunny spells. Meanwhile, a string of low-pressure systems waited in the wings. One of these was Storm Kathleen, a deep depression which arrived on the 5th. As Storm Kathleen approached from the west, it introduced a strong southerly flow all the way from Iberia, bringing mild conditions and temperatures up to 20.9 °C at Santon Downham in Suffolk. Upon its arrival, Storm Kathleen brought strong winds and heavy rain to northern and western areas, falling as snow over the mountains. Large waves swept a car into the sea in Fife, eastern Scotland, while strong winds damaged the roof of the Titanic Belfast centre. Storm Kathleen had barely cleared northern Scotland when the next storm arrived, named Storm Pierrick by the French meteorological service, Météo-France. This storm mainly affected the south coast of England and the Channel Islands, where it coincided with high spring tides, leading to coastal flooding. Bracklesham in West Sussex experienced such severe flooding that around 100 people were injured and a major rescue operation was launched.

Further low-pressure systems continued to bring unsettled conditions until around the 20th, when high pressure built to the west. It introduced a northerly airflow, bringing lower temperatures to the UK. It also brought a brief respite in the form of settled weather, before the next fronts began to arrive on the 25th. The return of low pressure brought a showery end to April.

The cool end to the month was balanced out by a mild start, leading to close-to-average temperatures across the UK for April as a whole, only 0.4 °C above average. Thanks to a persistent parade of low-pressure systems from the west bringing plenty of rain, April continued the trend of wetter-than-average months, with 155 per cent of its usual rainfall. It was also another dull month, with 79 per cent of average sunshine.

The averages are:

Maximum temperature	12.03 °C
Minimum temperature	3.76 °C
Rainfall	71.71 mm
Sunshine	155.32 hrs
Air frost days	4.03

Historic Events

1 April 1994 – One of the most intense April storms since records began brought a gust of 100 mph to Cardiff weather centre on Good Friday.

2 April 1917 – During the coldest April on record, temperatures fell as low as −15.4 °C, the lowest ever measured in April. The exceptional low was recorded at the Eskdalemuir weather station in Dumfriesshire, Scotland.

18 April 1849 – A severe snowstorm with bitterly cold northeasterly winds hit southern England. Writing from Buckingham Palace, Queen Victoria described the day as 'dreadful', adding: 'there were incessant showers of rain and snow, so cold. It was impossible to attempt to go out.'

23–26 April 1908 – Blizzards raged across southern Britain, with 30–40 cm accumulating in some places. On the 25th, falling snow made visibility dangerously low at sea, leading to a collision between HMS *Gladiator* and the SS *St Paul* off the coast of the Isle of Wight, with the *Gladiator* sinking.

30 April 1054 – The earliest tornado known in Europe hit Rosdalla, near the Irish town of Kilbeggan, County Westmeath. It's described in one account as a 'tower of fire' surrounded by 'birds' which picked up objects from the ground.

APRIL

In this month...

2 April – Pink Moon

17 April – New Moon

22 April – Earth Day

Look out for: Rainbows
With April showers, inevitably come rainbows. Seen when the Sun shines from behind the observer, they form as sunlight hits raindrops in front of the viewer. The light reflects and refracts, splitting into a spectrum of rainbow colours, with red on the outer arc and violet hues on the inside. Sometimes, sunlight is reflected twice within a raindrop, leading to a double rainbow. Wider than the primary rainbow, its colour scale is reversed and usually appears less bright.

A rainbow over high moorland.

Observing the Weather – Tracking a Passing Front

There's no better way to understand how weather fronts work than to watch them in action. The passage of a warm front followed by a cold front is accompanied by different clouds at each stage, so if you know what to look for, you'll be able to tell when the fronts have moved through. The first sign of an approaching warm front is high-level cirrus clouds – wispy, feather-like streaks – slowly spreading across the skies. As the front approaches, the clouds will get progressively lower, so cirrus will be followed by altostratus and eventually nimbostratus, which bring rain and the boundary of the warm front.

The moist and mild air that follows is known as the warm sector, the area between a warm front and a cold front. Rain will begin to ease, but it usually remains cloudy. However, the approaching cold front brings a return to wetter conditions, marked by a narrow band of heavy rainfall. After that, much brighter, cooler weather follows. However, the cold air creates instability and after a time, cumulus clouds may begin to form, developing into towering cumulonimbus and bringing heavy downpours. Keeping an eye on the Met Office forecast surface pressure charts can give you a heads-up when a frontal system is on its way. For more information on each of the various cloud types, see pages 234–245.

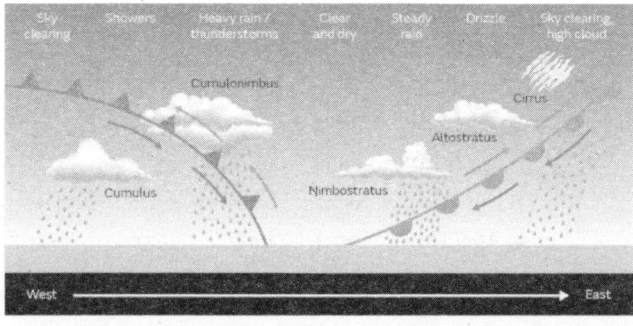

The sequence of clouds associated with the passage of a frontal system.

APRIL

Sun and Moon Times in April 2026

Location	Date	Sunrise	Sunset	Moonrise	Moonset
Belfast					
	01 Apr (Wed)	06:56	20:00	19:43	06:33
	11 Apr (Sat)	06:31	20:19	04:51	12:25
	21 Apr (Tue)	06:07	20:38	07:44	01:55
	30 Apr (Thu)	05:47	20:55	20:05	04:57
Cardiff					
	01 Apr (Wed)	06:48	19:45	19:27	06:23
	11 Apr (Sat)	06:25	20:02	04:22	12:31
	21 Apr (Tue)	06:04	20:19	07:59	01:16
	30 Apr (Thu)	05:46	20:34	19:43	04:54
Edinburgh					
	01 Apr (Wed)	06:43	19:50	19:33	06:21
	11 Apr (Sat)	06:17	20:11	04:49	12:05
	21 Apr (Tue)	05:52	20:31	07:18	01:57
	30 Apr (Thu)	05:31	20:49	19:59	04:43
London					
	01 Apr (Wed)	06:36	19:33	19:14	06:11
	11 Apr (Sat)	06:13	19:50	04:10	12:18
	21 Apr (Tue)	05:52	20:07	07:47	01:03
	30 Apr (Thu)	05:34	20:22	19:30	04:42

*all times in BST

Twilight Times in April 2026

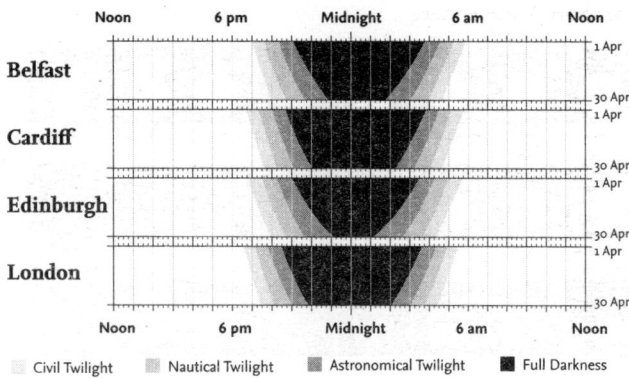

Thanks to the clocks changing at the end of last month, daylight hours through April begin to feel noticeably longer. From the start of the month to the end, sunrise becomes around an hour earlier, while the Sun sets almost an hour later, so about two hours of daylight are gained, depending on latitude. The increased likelihood of clearer skies in April gives ample opportunity to enjoy the lengthening daylight hours.

Moon Phases in April 2026

Northern Hemisphere

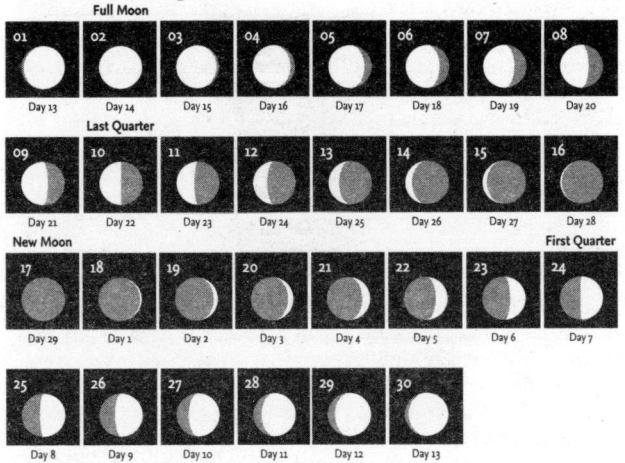

The Pink Moon
April's Full Moon is named after the blush hues of wildflowers coming into bloom at this time of year. It's thought that the 'Pink Moon' was named by Native Americans after the phlox flower, a native plant to North America which blossoms around April. In Europe, similar spring-like names are used, including the 'Budding Moon', the 'Egg Moon' and the 'Growing Moon'.

The Chernobyl Disaster

In the early hours of Saturday, 26 April 1986, the explosion of a reactor at the Chernobyl Power Plant led to the worst nuclear disaster in history. It happened after a routine safety test went tragically wrong, releasing colossal amounts of radioactive material into the atmosphere – around 400 times the amount unleashed by the atomic bomb dropped on Hiroshima. The worst-affected areas were those closest to the explosion in northern Ukraine. Approximately 150,000 square kilometres of Ukraine, Belarus and Russia were contaminated by radioactive fallout. In the aftermath of the disaster, the plume of radioactive material soon began to spread across Europe. With high pressure over northwestern Russia and low pressure across western Europe, resultant north to northeasterly winds carried radioactive particles all the way to Sweden in as little as two days. The spread of contamination was highly dependent on weather conditions, not only wind direction, but rain too, which dragged radioactive material out of the clouds and down to the ground. In the days that followed the disaster, it seems that Russian officials used this knowledge to their advantage, harnessing the power of the weather to prevent radioactive particles from reaching Moscow. It is widely believed that the Soviet Minister of Hydrometeorology, Yuri Izrael, ordered a mass cloud seeding operation, manipulating the clouds to make it rain. He sent military pilots into the skies above Belarus, where they dispersed jets of silver iodide. This causes water vapour to condense more readily, falling as rain. Following this, heavy, black-coloured rain was reported to have fallen across the city of Gomel, Belarus. Scientists suggested that Belarusians living in the affected areas were exposed to radiation doses 20 to 30 times higher than normal.

Elsewhere, clouds of radioactive material spread across the rest of the continent and every country in mainland Europe recorded above-average radiation. In the UK, radioactive tests detected an increase over high-ground areas such as the Scottish mountains, Cumbria and the Welsh mountains. A ban on the sale of livestock was put in place to prevent people from ingesting radiation from contaminated meat.

The disaster highlighted a major shortfall in forecasting the spread of airborne contaminants. In the years that followed,

APRIL

meteorological organizations across the world began to develop their own dispersion models. In 1993, the UK Met Office launched their Numerical Atmospheric-dispersion Modelling Environment (NAME), capable of simulating a wide range of events, from nuclear disasters like Chernobyl to volcanic events such as the 2010 eruptions of Icelandic volcano Eyjafjallajökull.

Mandatory testing and tagging of sheep presisted for over twenty years in some parts of Scotland, Wales and Cumbria. Restrictions were finally lifted in 2010 for Scotland, and 2012 for England and Wales.

Introduction

Blue skies, green grass and gardens a riot of colour illuminated by May sunshine – the last month of meteorological spring is here. However, many cultures start the month celebrating the arrival of summer, with European May Day and the Celtic festival of Beltane both marking the beginning of warmer, sunnier and drier months. May does seem an apt place to start, as the sunniest and driest month on average in the UK. With blocking highs most likely to dominate our weather during this month, clear skies bring an abundance of late spring sunshine. Despite its settled reputation, May gets its fair share of severe weather. Even when things appear warm and fine inland, coastal areas can tell a very different story. Warm air passing over the relatively cold North Sea cools and condenses, forming a thick coastal fog common along the eastern shores of the UK. Depending on where you are, it's known by different names: haar in eastern Scotland and fret in eastern England. Also known as advection fog (see page 89), it's most likely to occur in late spring and summer, when our weather is generally warming up, but the sea is relatively cold.

Thunderstorms are also more likely through May and the following summer months. They require atmospheric instability, which causes air to rise, cool and condense. By this process, towering cumulonimbus clouds can form. Inside these clouds, electric charge can build, leading to a lightning strike and a crash of thunder. May has a long record of severe thunderstorms and the Hertfordshire hailstorm of 1697 is perhaps one of the most notable. Classified as a destructive 'H8' hailstorm by The Tornado and Storm Research Organisation (TORRO) (see page 258), it was the most intense hailstorm ever recorded in Britain. Tracking around 25 km from Hitchin, Hertfordshire, to Potton in Bedfordshire on 15 May, hailstones up to 11 cm in diameter were recorded, with some reports suggesting pieces of ice of 14 cm. Caught in the open when the storm hit, a young man tending sheep was killed.

Even the warmest May day on record, in 1944, saw violent thunderstorms. Several weather stations across London and southeastern England recorded a high of 32.8 °C on the hot and sunny afternoon of 29 May. Conditions soon deteriorated though, with thunderstorms bringing frequent lightning, large hailstones and severe flash flooding which led to several deaths in a town in Yorkshire.

Only six years later another historical weather event took place. On 21 May 1950, a famously long and destructive tornado wrought havoc as it tracked 65 miles across the east of England. Born from thunderstorms triggered by a plume of warm, humid air from France, the tornado touched ground near Wendover in Buckinghamshire, creating a path of destruction as it moved northeastwards, eventually lifting into a funnel cloud near Ely in Cambridgeshire. The same funnel cloud was shortly spotted in Shipham in Norfolk. It lasted two and a half hours and was accompanied by large hail and torrential rain, leading to further carnage. Homes were damaged, trees were uprooted and a double-decker bus overturned near Ely. A little girl was killed in flash flooding while some reports suggest two men also lost their lives during the severe weather.

Weather word: Fret (Northern English, eighteenth century)
A wet mist, fog or haze coming inland from the sea.

Weather Extremes in May

Country	Temp.	Location	Date
Maximum temperature			
England	32.8 °C	Camden Square (London)	22 May 1922
		Horsham (West Sussex)	29 May 1944
		Tunbridge Wells (Kent)	29 May 1944
		Regent's Park (London)	29 May 1944
Wales	30.6 °C	Newport (Monmouthshire)	29 May 1944
Scotland	30.9 °C	Inverailort (Inverness-shire)	25 May 2012
Northern Ireland	28.3 °C	Lisburn (County Antrim)	31 May 1922
Minimum temperature			
England	−9.4 °C	Lynford (Norfolk)	4 May 1941
			11 May 1941
Wales	−6.2 °C	St Harmon (Powys)	14 May 2020
Scotland	−8.8 °C	Braemar (Aberdeenshire)	1 May 1927
Northern Ireland	−6.5 °C	Moydamlaght (County Londonderry)	7 May 1982

Country	Pressure	Location	Date
Maximum pressure			
Republic of Ireland	1042.4 hPa	Cork Airport (County Cork)	12 May 2012
Minimum pressure			
England	968.0 hPa	Sealand (Flintshire)	8 May 1943

MAY

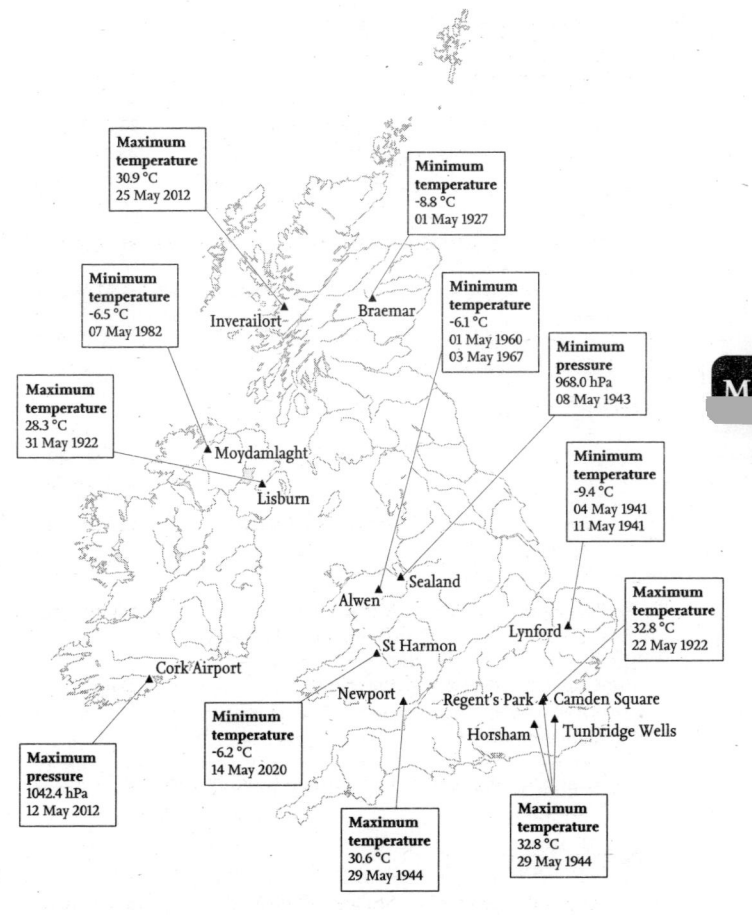

The Weather in May 2024

Observation	Location	Date
Max. temperature 27.5 °C	Chertsey, Abbey Mead (Surrey)	12 May
Min. temperature −1.1 °C	Kinbrace Hatchery (Sutherland)	21 May
Most rainfall 124.0 mm	Honister Pass (Cumbria)	22 May
Most sunshine 15.7 hrs	Lerwick (Shetland)	16 May
Highest gust 55 mph (88 kph/48 kt)	Orlock Head (Down)	23 May

Although some late frosts caught gardeners out in the first few days, temperatures were generally well above average, resulting in the warmest May on record. It began as April left off, with frequent spring showers, some of these heavy and thundery. Frequent lightning damaged homes and power supplies in Sussex and Wiltshire, while flooding affected southwestern England, Leeds and Aberdeenshire.

After about a week, high pressure began to build, bringing a spell of warm, settled weather. Clear skies gave people across the UK a glimpse of the aurora borealis at night, during the strongest geomagnetic storm since 2003. Temperatures climbed into the high 20s, peaking at 27.5 °C on 12 May in Surrey. Following the height of the warmth, thunderstorms spread across Northern Ireland, Wales, Scotland, the West Midlands and northern England. Many areas saw surface-water flooding, and marble-sized hailstones were reported in Gloucestershire. The torrential downpours even disrupted a football match, as the roof of the stadium began to leak towards the end of the Manchester United vs Arsenal game – known infamously as the 'Old Trafford waterfall'. The thunderstorms marked a fall in temperatures, which returned to a little closer to average, as well as a shift in the weather pattern. Thereafter, May remained unsettled with only occasional drier and brighter interludes. From the 21st, a depression tracked northwards from central Europe towards eastern Scotland, deepening as it moved over the North Sea. It brought persistent rain and disrupted rail services up and down the country, while some houses in Cumbria were cut off by flooding.

May was another wet month, with 116 per cent of average rainfall overall, although the localized nature of thunderstorms meant rainfall totals were quite variable. Temperatures were a record-breaking 2.4 °C above average, while sunshine was once again below average, at 83 per cent. It was also the warmest spring on record, mainly thanks to May at the end, and wetter than average for most places too. Only northwestern Scotland was drier than average, where instead of rain, wildfires blazed in the warm and windy spring weather. This was in stark contrast to some eastern counties, for which it was their wettest spring on record.

The averages are:

Maximum temperature	15.13 °C
Minimum temperature	6.25 °C
Rainfall	70.96 mm
Sunshine	192.15 hrs
Air frost days	1.2

Historic Events

4 May 1941 – Temperatures plummeted to a breathtaking −9.4 °C at Lynford near Thetford Forest in Norfolk thanks to cold northerly winds. It remains the lowest temperature recorded in May but was equalled just a week later as temperatures once again fell below −9 °C at the same site.

18 May 1891 – Many awoke on the spring bank holiday to a covering of snow following flurries the previous day and overnight. The weather in London was described as 'raw and cold' but becoming 'somewhat finer' later in the day as highs rose to 6.1 °C at Greenwich.

19 May 1989 – On a hot and humid Friday afternoon, large black clouds started to form over the hills above Hebden Bridge, West Yorkshire. Within minutes, torrential rain began to fall. Homes across Calderdale were flooded and roads turned to rivers as water levels rapidly rose. Reports suggest 193 mm of rain fell within two hours during the event, over a very localized area.

23 May 2011 – A deep depression brought widespread gales to northern Britain with gusts up to 100 mph. The Forth and Tay road bridges were closed, power cuts affected thousands of people across Scotland and trees, in full spring leaf, were uprooted.

29 May 1920 – 23 people lost their lives during a catastrophic flash flood in Louth, a town in Lincolnshire. A shallow depression spreading into southwestern England brought widespread thunderstorms, with reports suggesting more than 100 mm of rain fell within three hours over the Lincolnshire

MAY

Wolds. Water levels of the River Lud rose very suddenly by up to two metres, the deluge sweeping people off their feet and destroying houses.

In this month...

1 May – Flower Moon

16 May – New Moon

31 May – Blue Moon

Look out for: Advection fog
This type of fog develops when warm air is moved, or 'advected', over a cooler surface. It causes water vapour in the air near the surface to cool and condense into tiny droplets of water, forming fog. As mentioned on page 82, it can happen when relatively warm air moves over cool seas, but can also occur when a warm front passes over ground blanketed in snow.

Observing the Weather – Sun Dogs

Towards the end of the day, when the Sun is low in a sky painted with cirrus clouds, it's worth looking out for optical phenomena developing near the Sun. Sun dogs, also known as mock suns or, to give them their scientific name, parhelia, appear as bright, rainbow-like spots either side of the Sun around sunset or sunrise. They're caused by refraction of sunlight through the ice crystals present in cirrus clouds high in the sky, especially cirrostratus. The process of refraction splits the light into a range of rainbow colours, with the blue end of the spectrum towards the outside of the sun dogs and the red hues towards the inner edges nearest the Sun. The two sun dogs are not always equal; sometimes one is brighter than the other or one does not appear at all.

Sun dogs form when ice crystals within high clouds are flat and horizontally aligned. If they're more randomly shaped and aligned, then sunlight will generally be refracted to form a halo. Apparent distances in the sky can be measured in degrees by imagining a giant protractor stretched from left to right (or in front to behind), with you at its centre. By this system, sun dogs and halos appear at a distance of 22° from the Sun, which is why they're sometimes called 22° halos.

Parhelia at sunrise.

Sun and Moon Times in May 2026

Location	Date	Sunrise	Sunset	Moonrise	Moonset
Belfast					
	01 May (Fri)	05:45	20:57	21:25	05:07
	11 May (Mon)	05:25	21:16	03:32	14:11
	21 May (Thu)	05:08	21:33	09:36	02:03
	31 May (Sun)	04:55	21:47	22:55	04:08
Cardiff					
	01 May (Fri)	05:44	20:35	20:58	05:08
	11 May (Mon)	06:27	20:51	03:14	14:04
	21 May (Thu)	05:12	21:06	09:42	01:33
	31 May (Sun)	05:02	21:19	22:17	04:20
Edinburgh					
	01 May (Fri)	05:29	20:51	21:21	04:50
	11 May (Mon)	05:08	21:11	03:24	13:57
	21 May (Thu)	04:50	21:29	09:18	02:01
	31 May (Sun)	04:36	21:45	22:57	03:45
London					
	01 May (Fri)	05:32	20:23	20:45	04:55
	11 May (Mon)	05:15	20:39	03:02	13:51
	21 May (Thu)	05:00	20:54	09:29	01:21
	31 May (Sun)	04:49	21:07	22:04	04:08

*all times in BST

Twilight Times in May 2026

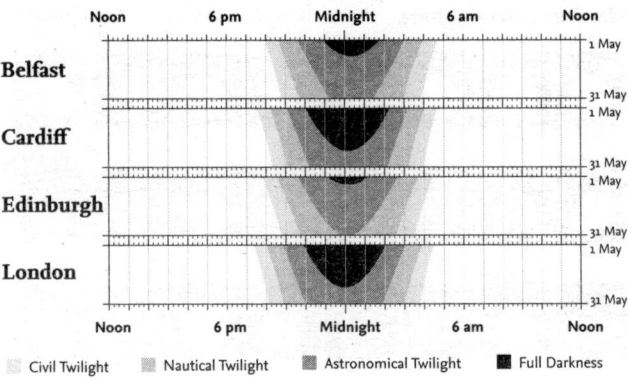

The sunniest month of the year on average, May's lengthening daylight hours give plenty of opportunity to enjoy the settled weather, should it prevail. At some point during the month, depending on your latitude, true darkness ceases to exist and instead the night-time passes in a state of twilight, as the Sun no longer dips more than 18 degrees below the horizon. In southern Cornwall, this doesn't happen until the last day of the month, while to the north of Aberdeen, this actually occurs in the last few days of April.

Moon Phases in May 2026

Northern Hemisphere

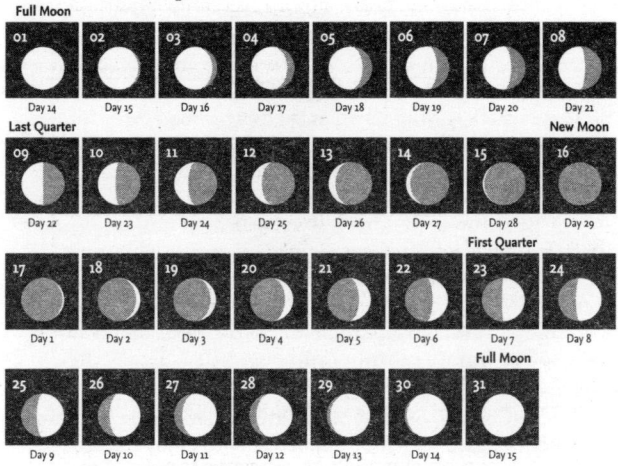

The Flower Moon
You don't have to look far for the inspiration behind the naming of May's Full Moon. The 'Flower Moon' represents the bloom taking place at this time of year. The alternative Anglo-Saxon name of 'Milk Moon' refers to how often cows were milked during May – three times a day according to the Old English word 'Rimilcemona' – roughly translating to 'the month of three milkings'. Alternative Celtic and Old English names for the Full Moon in May are: 'Hare Moon', 'Grass Moon', 'Mothers' Moon' and 'Bright Moon'.

In May 2026, a relatively rare Blue Moon will appear on the last day of the month. It has nothing to do with the colour of the Moon, instead referring to the fact that it's the second Full Moon of the month. This happens once every two or three years.

Daylight Saving Time

Twice a year billions of people around the world pause for a moment to change their clocks and adjust for Daylight Saving Time. The popular phrase 'spring forward and fall back' reminds us that in spring, time shifts an hour forwards, meaning we lose an hour of sleep, while in autumn or fall, time moves back, giving us an extra hour in bed. Around 70 countries worldwide observe Daylight Saving Time, a method introduced to make the most of the long hours of daylight during summer. Here, it's known as British Summer Time, or BST. It was first proposed by British builder William Willett, from Chislehurst in southeast London. The story goes that he was riding his horse one summer morning and noticed rows upon rows of houses with their curtains still drawn tightly shut, despite the Sun having been up for hours. This inspired him to create a campaign titled 'The Waste of Daylight'. His idea was to move the clocks forward by 20 minutes on each of the four weekends in April, and reverse the change over four weekends in September. He wrote it all down in a pamphlet, which he self-published in 1907. He gathered many supporters, including Winston Churchill and Arthur Conan Doyle, but the first to suggest it to the House of Commons was MP Robert Pearce. However, the Daylight Saving Bill was defeated in 1909 and Mr Willett never lived to see his idea come to fruition, dying in 1915. A year later, as the First World War raged, the idea was revisited as a means of reducing demand on coal. On 17 May 1916 the Summer Time Act was passed and the following Sunday, 21 May, the clocks went forward by an hour for the first time. The clocks were put back to GMT on 1 October. And so it remained this way for the next 20 years or so, until the Second World War came along. Instead of putting the clocks back at the end of British Summer Time in 1940, they stayed as they were, so when the clocks went forward again in spring 1941, the UK was now two hours ahead of GMT and British Double Summer Time was born. The theory was that the two hours of extra daylight in the evening would give people longer to get home before the blackout. Once the war was over, the UK returned to GMT and BST. In the current arrangement, British Summer Time begins on the last Sunday of March and ends on the last Sunday of October. It's still a topic of great debate, with many attempts being made to change it over the years.

William Willett.

Introduction

With meteorological summer beginning on 1 June and astronomical summer arriving later in the month, you'd be forgiven for expecting warm, sunny weather, perfect for sunbathing in the garden or a quick dip in the gradually warming sea. And often, June lives up to our expectations. The year 1925 takes the top spot for the driest June on record, with plenty of sunshine recorded too. Weather stations at Falmouth, Ross-on-Wye and Calshot failed to measure a single millimetre of rain, while at Kew in London only a measly one millimetre was recorded. Meanwhile, sunshine hours soared with Falmouth breaking its previous record of June sunshine by nearly 30 hours, basking in a total of 375 hours of sunshine through the month. The pressure pattern that led to this remarkably dry and sunny weather was split between a persistent anticyclone and pleasant warmth for half of the month and cool northerlies for the other half. Balancing out, this combination led to around average temperatures for the month as a whole.

The summer solstice on 21 June brings the longest day of the year and marks the beginning of astronomical summer. Naturally, the days leading up to and away from the solstice are also long. While the month often exhibits blue skies and wall-to-wall sunshine as in 1925, it is not the sunniest on average, that title falling to May. In fact, May 2020 was the sunniest of any month on record with almost 270 hours of sunshine, while June 2020 recorded only 161.2 hours. This is thanks to a phenomenon known as the 'return of the westerlies' which occurs in the first few weeks of June. It marks the end of the period in late spring when anticyclonic blocking and winds from the northeast become more likely. As the anticyclones dissipate, Atlantic depressions are once again able to resume their eastward track to the north of the UK and thus the predominant wind direction returns to a familiar westerly. Therefore, as June proceeds, skies are likely to be much more dynamic than those in May, thanks to weather fronts bringing a parade of different cloud types as they traverse the country. However, these low-pressure systems are no longer the wild winter storms driven by a powerful jet stream, they are much shallower and don't tend to bring as much in the way of severe weather. This is because the temperature contrast between the Equator and the poles is much lower in the summer, so the jet stream is nowhere near as strong or capable of whipping up a deep extratropical cyclone.

However, a little bit of rain will come as a relief to hayfever sufferers at this time of year. June marks the peak in the grass pollen season, the type of pollen that most commonly causes allergies. Rain helps to clear the air and suppress the pollen count, while the worst weather for allergy sufferers is warm, dry and breezy.

Weather word: Thunderplump (English, nineteenth century)
A sudden torrential downpour that soaks you to the skin in seconds, accompanied by thunder and lightning.

Weather Extremes in June

Country	Temp.	Location	Date
Maximum temperature			
England	35.6 °C	Camden Square (London)	29 June 1957
		Southampton Mayflower Park	28 June 1976
Wales	33.7 °C	Machynlleth (Powys)	18 June 2000
Scotland	32.2 °C	Ochtertyre (Perth and Kinross)	18 June 1893
Northern Ireland	30.8 °C	Knockarevan (Country Fermanagh)	30 June 1976
Minimum temperature			
England	−5.6 °C	Santon Donwham (Norfolk)	1 June 1962 / 3 June 1962
Wales	−4.0 °C	St Harmon (Powys)	8 June 1985
Scotland	−5.6 °C	Dalwhinnie (Inverness-shire)	9 June 1955
Northern Ireland	−2.4 °C	Lough Navar Forest (County Fermanagh)	4 June 1991

Country	Pressure	Location	Date
Maximum pressure			
Republic of Ireland	1043.1 hPa	Clones (County Monaghan)	14 June 1959
Minimum pressure			
Scotland	968.4 hPa	Lerwick (Shetland)	28 June 1938

JUNE

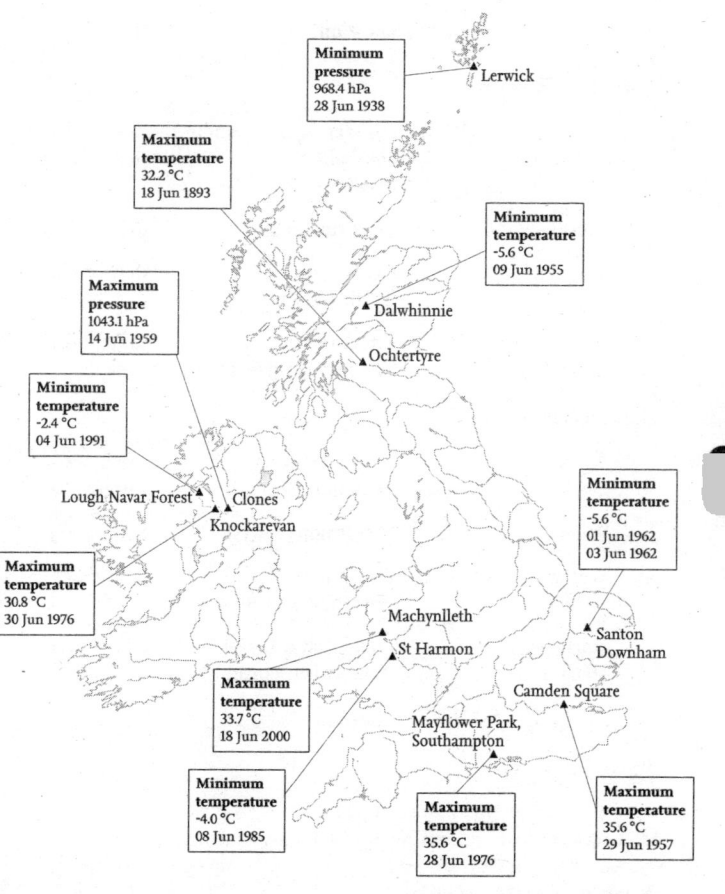

WEATHER ALMANAC 2026

The Weather in June 2024

Observation	Location	Date
Max. temperature 30.5 °C	Wisley (Surrey)	26 June
Min. temperature −1.6 °C	Kinbrace, Hatchery (Sutherland)	13 June
Most rainfall 51.2 mm	Mickleden, Middlefell Farm (Cumbria)	13 June
Most sunshine 16.1 hrs	Dyce (Aberdeenshire)	24 June
Highest gust 63 mph (101 kph/55 kt)	Wight: Needles Old Battery (Isle of Wight)	15 June

Following a record-breakingly warm May, June was notably cool, particularly in the first two weeks. The month began with high pressure centred to the west of Ireland, the resultant north to northwesterly flow bringing Arctic air and a very fresh feel to conditions, with maximum temperatures scarcely making it higher than the mid-teens. Rather than relenting as June progressed, the chilly northerly airflow strengthened into the second week, as low pressure across Scandinavia increased the number of isobars across the country. By the middle of June, the mean temperature for the UK was more than 2 °C below average with little in the way of rain, either. However, an Atlantic depression arriving on the 13th/14th put an end to the dry spell, stalling across the UK and bringing 25–50 mm of rain to some parts of Ireland, Wales and western England. As the system cleared, thunderstorms developing along a convergence line brought heavy rainfall and surface-water flooding to parts of West Yorkshire on the 18th. After that, high pressure began to build again and temperatures finally recovered to what would be expected at this time of year. Skies cleared and strong sunshine allowed daytime highs to climb. By the 20th, highs

JUNE

were in the low-to-mid 20s. Coupled with the dry conditions of the month so far, it likely contributed to a wildfire that broke out on heathlands near Exmouth in Devon. Temperatures continued to rise over the next few days, peaking on the 26th in the southeast, when a high of 30.5 °C was recorded in Surrey. However, a deep low-pressure system arrived in the north on the 27th, bringing a cooler and more unsettled end to the month for most places.

The brief warm spell did almost enough to even out the cold start to the month, with the resultant mean temperatures only 0.4 °C below average. Meanwhile, with high pressure never too far away, it was also drier than average, with the UK only receiving 71 per cent of average rainfall for June. At 104 per cent, sunshine hours were close to average.

The averages are:

Maximum temperature	17.68 °C
Minimum temperature	9.08 °C
Rainfall	77.19 mm
Sunshine	171.49 hrs
Air frost days	0.07

Historic Events

2 June 1975 – Widespread snow brought several county cricket matches to a standstill across the Midlands and East Anglia. Flurries of snow were even reported as far south as Lord's Cricket Ground in London, with sleet falling in Portsmouth.

5 June 1983 – A large spider crab fell from the sky moments before a severe hailstorm began in Brighton. The crab, measuring nearly 25 cm across, appeared to drop out of a dark cumulonimbus cloud. It was followed by hailstones the size of marbles and accompanied by waterspouts developing offshore. Further storms hit up and down the coast of southern England, with hail as large as five centimetres in places.

13 June 2023 – Until 2023, 13 June was the only day of meteorological summer to have never recorded an official maximum temperature of 30 °C since records began. This statistical quirk was known as the '13 June enigma' until a high of 30.8 °C was recorded at Porthmadog, Wales, in 2023.

28 June 2018 – A high of 33.2 °C was recorded at Motherwell, near Glasgow, as a late June heatwave brought widespread warmth to the UK. At the time, it was considered to be Scotland's highest temperature on record, beating the previous record of 32.9 °C set in August 2003. However, upon investigation by the Met Office it was found that a stationary vehicle was parked nearby with its engine running and the record was rejected.

JUNE

30 June 1908 – An asteroid tumbled into Earth's atmosphere and exploded in a remote part of Siberia, flattening an area of trees larger than London. Seismic shockwaves from the event were registered in England and people reported seeing glowing skies and spectacular sunsets, thanks to the dust and ice particles dispersed into the atmosphere.

In this month...

1 June – Meteorological summer begins

8 June – World Ocean Day

15 June – New Moon

21 June – #ShowYourStripes Day

21 June – Astronomical summer begins

21 June – Summer Solstice

30 June – Strawberry Moon

Look out for: Dust devils
A dust devil, also known as a 'willy willy', is a swirling vortex of air. It may be similar in appearance to a tornado, but forms in a completely different way and is much less destructive. Whereas tornadoes reach down from clouds, dust devils grow upwards from the ground. They form when the ground is hot and dry, allowing strong updrafts to develop. Picking up dust, the air rises and rotates, becoming faster as the column of air stretches. They usually die out after a few minutes but if hot air near the surface persists, so can the dust devil.

Observing the Weather – Making a Simple Sundial

With the Sun high in the sky and plenty of daylight, the summer solstice is the perfect day for making a sundial – provided you have clear skies! The earliest sundials provided the first means to tell the time, indicated by the position of a shadow cast by an object illuminated by the Sun. To see this principle for yourself, you can make a simple sundial in your own garden or another outdoor space.

You will need:

- A straight stick – a bamboo cane or metre ruler work well
- 12 stones

Starting just before 7 a.m, plant your stick somewhere you know will see the Sun all day, and slant it slightly northwards. If you don't have soft ground or you're in a shared space, planting the stick in a bucket of sand will work too. When the clock hits 7 a.m., take a look at where the shadow of your stick has been cast and place a stone there. Return every hour and repeat the process (you could set an alarm just before each hour so you don't forget). At the end of the day your simple sundial will be finished and, provided the Sun is shining, you can use it to tell the time.

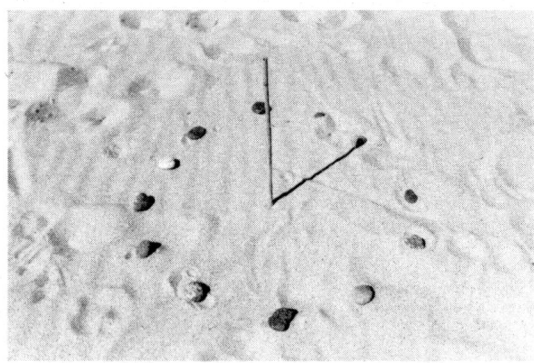

A simple sundial in the sand.

JUNE

Sun and Moon Times in June 2026

Location	Date	Sunrise	Sunset	Moonrise	Moonset
Belfast					
	01 Jun (Mon)	04:54	21:48	23:50	04:44
	11 Jun (Thu)	04:48	21:59	02:13	17:35
	21 Jun (Sun)	04:47	22:03	13:04	00:59
	30 Jun (Tue)	04:51	22:03	23:01	04:31
Cardiff					
	01 Jun (Mon)	05:01	21:20	23:12	04:59
	11 Jun (Thu)	04:55	21:29	02:10	19:11
	21 Jun (Sun)	04:55	21:33	12:53	00:46
	30 Jun (Tue)	04:59	21:33	22:29	04:45
Edinburgh					
	01 Jun (Mon)	04:35	21:46	23:53	04:19
	11 Jun (Thu)	04:27	21:57	01:58	17:29
	21 Jun (Sun)	04:26	22:02	12:53	00:49
	30 Jun (Tue)	04:30	22:01	23:01	04:06
London					
	01 Jun (Mon)	04:49	21:08	22:59	04:46
	11 Jun (Thu)	04:43	21:17	01:58	16:59
	21 Jun (Sun)	04:43	21:21	12:40	00:34
	30 Jun (Tue)	04:46	21:21	22:16	04:32

*all times in BST

Twilight Times in June 2026

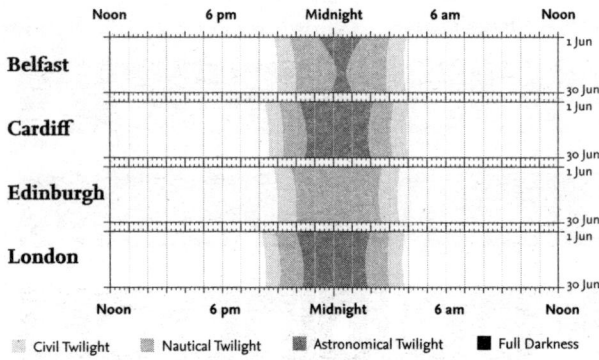

Throughout June, the nights exist in a state of twilight, never full darkness. For most of England and Wales, the Sun dips far enough below the horizon for astronomical twilight. However, for northern England, parts of Northern Ireland and Scotland, they only have civil and nautical twilight. This is most pronounced on the summer solstice on 21 June, the longest day of the year. In northern Scotland there are around 18 hours of daylight while in southwestern England there are just under 16.5.

Moon Phases in June 2026

Northern Hemisphere

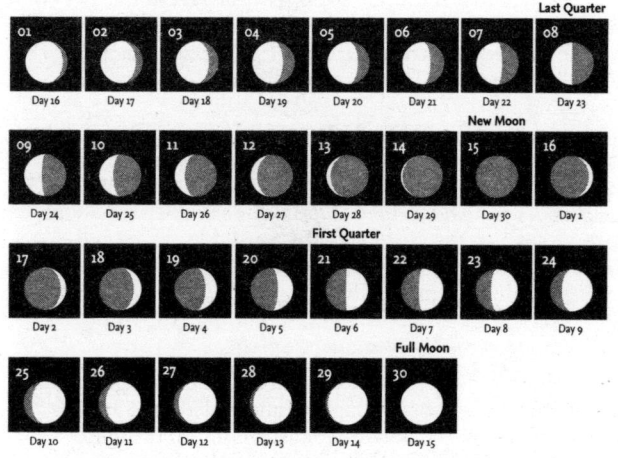

The Strawberry Moon
After months of planting and tending, gardeners begin to see the fruits of their labour as summer begins. June's 'Strawberry Moon' celebrates this too, named after the ripening and beginning of the berries' harvest. Other names for this month's Full Moon also mark the changes taking place at this time of year. The 'Rose Moon' is associated with the blooming of these flowers in June, while the 'Mead Moon' is named after the drink made from honey and grains. Meanwhile in Eastern cultures, it is sometimes called the 'Lotus Moon', after the flower blooming in ponds in summer.

Climate Warming Stripes

Most people living in the UK have experienced extreme weather. Whether it was the summer heatwave of 2022 or one of the many instances of flooding from winter storms, these events have been exacerbated by climate change. Yet it's still hard to visualize what that all means in terms of how things have changed. That's where #ShowYourStripes comes in. Every year on 21 June – a date which often marks the summer solstice – a powerful image is shared. Projected onto buildings, bridges and even the White Cliffs of Dover, a series of colourful stripes help spark conversations about climate change. Designed by Professor Ed Hawkins and debuted at the Hay Festival in 2018, the 'Warming Stripes' show how global temperatures have changed since 1850. With no words and no numbers, the resultant picture is one that can be understood by everyone. Consisting of a barcode-like series of stripes, each line represents the average temperature for a single year, relative to the average temperature over the 50 years between 1961 and 2010. Below-average years appear as a blue stripe, while above-average are red. From left to right, the colours shift from varying shades of blue to deepest red, showing the rapid warming of our planet. Since their creation, the Warming Stripes have been around the world. Painted into murals, shown on television and incorporated into fashion, the stripes have been shared in so many ways. But their origin is a rather humble one. In 2017, Professor Ellie Highwood, former climate scientist and colleague of Professor Hawkins, posted a picture of a 'global warming blanket' on social media. With one comment describing it as the 'most frightening knitwear' they'd seen all year, Professor Highwood's crochet masterpiece featured 15 different colours of yarn to make stripes representing the warming climate. It was this that inspired Professor Hawkins to go on to create the Warming Stripes graphic, but it was not the first instance of a climate-inspired crafting project. Two years previously another scientist-turned-crocheter, Joan Sheldon, created a scarf using a similar principle. Using hues of blue and purple fading into red, the marine scientist was likely the first to visualize complex climate data in such a creative way.

JUNE

From their creative inception to becoming a global sensation, the Warming Stripes exist to raise awareness of how our climate is changing. Therefore, participating in #ShowYourStripes day couldn't be simpler. By visiting the Show Your Stripes website, you can download your local stripes and share using the hashtag on social media. Or you might be inspired by Ellie Highwood and Joan Sheldon to do something creative. Either way, by sharing your stripes and starting a conversation, you can spread a powerful message: our climate is changing and we demand action.

Professor Ed Hawkins.

Joan Sheldon's scarf.

Introduction

Long hazy days, stuffy nights and grassy meadows beginning to turn crisp and brown, the warmest month of the year is upon us. Not only is July the warmest month on average, it's also most likely to record the hottest day of the year. Records were broken in July 2022 when temperatures reached and exceeded 40 °C for the first time in the UK. Scientists say this couldn't have happened if not for human-induced climate change. The likelihood of extremely hot days is increasing too. By the end of the century, summers with a 40 °C day will occur every three or four years on average if greenhouse gas emissions remain high. If emissions are reduced, the chance of these extreme temperatures becomes significantly lower. Along with the record-breaking daytime temperatures in 2022, the UK also experienced several tropical nights, when the temperature didn't fall below 20 °C overnight. In this particular extreme heatwave, an Oxfordshire weather station recorded a night-time low of 26.8 °C – the warmest night on record. Temperatures of that magnitude are well above the average for the daytime, let alone at night. Heat like this is extremely uncomfortable to many, if not most, people. However, there are simple steps that can be taken to keep yourself and your home cooler. Although people are often tempted to throw open the windows in the hope of bringing in a cooling breeze, the best way to keep your house cool is by closing the windows and curtains during the day, so as not to let any of the heat in. The coolest time of day is usually in the early hours of the morning during summer, so this is the best time to open your windows for some fresh air. Meanwhile, the hottest part of the day tends to be between 11 a.m. and 3 p.m. This is when you should try to stay indoors or in the shade, avoid physical exertion and drink plenty of water.

Although heat such as that of July 2022 was unheard of before the twenty-first century, there were still plenty of hot days, and those falling between July and August have historically been referred to as 'dog days'. Despite the images of hot, lazy dogs lounging in sunshine that come to mind, the term is more to do with the stars. In particular, Sirius, the brightest star in the constellation Canis Major, known as the Dog Star. The dog days begin when Sirius appears to rise around the same time as the Sun in mid-July. Coinciding with what are often the warmest days of the year, the Romans believed that Sirius was partly responsible for the heat, theorizing that the star was so bright it must be contributing extra warmth. The dog days end around mid-August.

Despite its warming trend and association with the hottest days of the year, there have been some historically cold Julys too. July 1888 is the joint coldest on record with 1922, with an average temperature of 12.3 °C. On 11 July 1888, snow reportedly fell across the northwest Midlands, the Lake District and Scotland, while some accounts even suggest flurries in London! Although there are some doubts over the reliability of snow reports in London, many newspapers reported snow over high ground, including Skiddaw. The Met Office weather report issued at the time confirmed snow had fallen in places, and described the weather pattern as a 'strong, squally northerly wind of exceptionally low temperature for the time of year'.

Weather word: Swullocking (English, nineteenth century)
Old East Anglian dialect meaning very hot and humid. Derived from 'swelking', which means sultry.

Weather Extremes in July

Country	Temp.	Location	Date
Maximum temperature			
England	40.3 °C	Coningsby (Lincolnshire)	19 July 2022
Wales	37.1 °C	Hawarden Airport (Flintshire)	18 July 2022
Scotland	34.8 °C	Charterhall (Scottish Borders)	19 July 2022
Northern Ireland	31.3 °C	Castlederg (Country Tyrone)	21 July 2021
Minimum temperature			
England	−1.7 °C	Kielder Castle (Northumberland)	17 July 1965
Wales	−1.5 °C	St Harmon (Powys)	3 July 1984
Scotland	−2.5 °C	Lagganlia (Inverness-shire)	15 July 1977
Northern Ireland	−1.1 °C	Lislap Forest (County Tyrone)	17 July 1971

Country	Pressure	Location	Date
Maximum pressure			
Scotland	1039.2 hPa	Aboyne (Aberdeenshire)	16 July 1996
Minimum pressure			
Scotland	967.9 hPa	Sule Skerry (Northern Isles)	8 July 1964

JULY

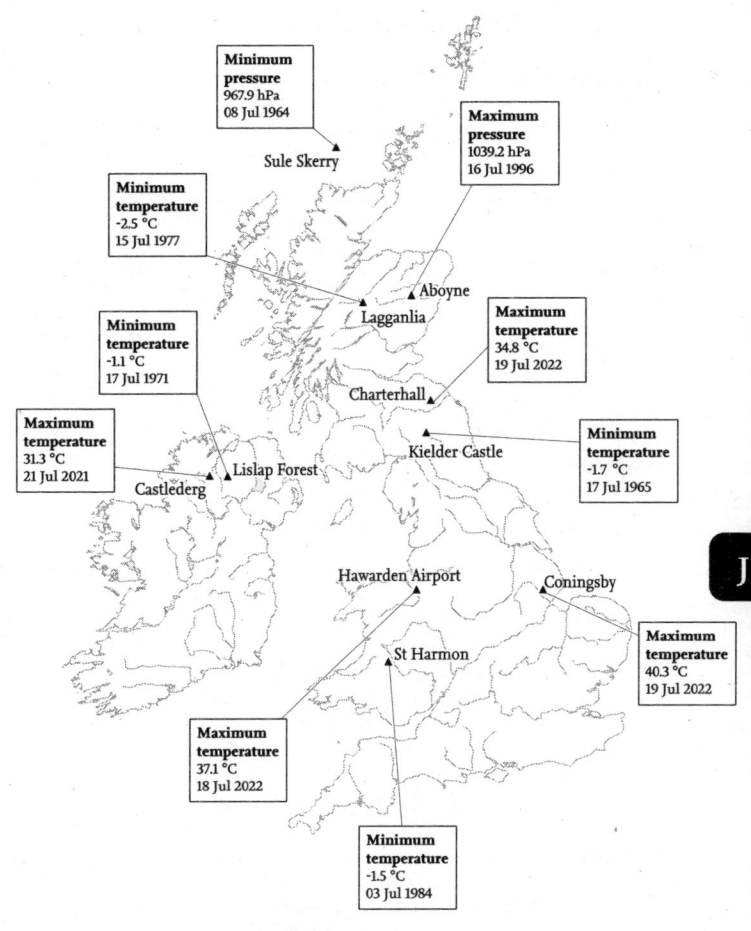

The Weather in July 2024

Observation	Location	Date
Max. temperature 32.0 °C	Heathrow (Greater London) and Kew Gardens (Greater London)	30 July
Min. temperature 1.7 °C	Tulloch Bridge (Inverness-shire) and Tyndrum No. 3 (Perthshire)	31 July
Most rainfall 75.7 mm	White Barrow (Devon)	8 July
Most sunshine 15.3 hrs	Dale Fort (Dyfed)	7 July
Highest gust 58 mph (93 kph/50 kt)	Brizlee Wood (Northumberland) Needles Old Battery (Isle of Wight)	4 July 5 July

July continued the cooler-than-average theme established in the first month of summer 2024, but where June was cool and dry, July was much more unsettled. This was thanks to a stronger-than-average jet stream situated much further south than usual, pushing a series of low-pressure systems across the UK. Conditions were changeable, with a particularly wet and windy day in the south on the 5th as a shallow depression tracked across southern England. The 8th and 9th of July saw a more widespread rainfall event as fronts spread northeastwards. In the southwest, 30 mm of rain fell widely, with more than 70 mm of rain over parts of Cornwall. As the system progressed, Cumbria and northeastern England were also hit, with rainfall totals of 40–50 mm.

By mid-month, parts of eastern and southwestern England had already recorded more than their average rainfall for July. Similar to the previous month, temperatures were 2 °C

JULY

below average by this point and it was the coldest start to July since 2004. From here, conditions tried to turn more settled. High pressure brought drier days around the 13th and again from the 16th. The second of these anticyclonic spells saw south-to-southwesterly winds bringing warm continental air. Temperatures began to climb for the first time this month, peaking on the 19th when highs of 30 °C were recorded widely across the southeast before the heat gradually slipped away. Over the following days further Atlantic systems brought scattered showers, longer spells of rain and closer-to-average temperatures. However, high pressure swooped in again towards the end of July, bringing a brief dry, settled and hot spell. This time, the heat was confined to a smaller area, reaching 32.0 °C in London on the 30th – the highest temperature of the summer so far.

The two hot spells during the second half of the month managed to bring the overall temperature for July much closer to average, only 0.5 °C below usual. Averaged across the whole of the UK, rainfall was normal for July, with wetter conditions in the south and east balanced out by drier weather in the northwest. Meanwhile it was rather dull, with only 89 per cent of average July sunshine.

The averages are:

Maximum temperature	19.62 °C
Minimum temperature	11.02 °C
Rainfall	82.46 mm
Sunshine	173.44 hrs

Historic Events

8 July 1841 – Alongside torrential rain and hailstones, hundreds of small fish and frogs were reported to have fallen from the sky during a heavy thunderstorm over Derby.

9 July 1959 – The first identified supercell thunderstorm hit Wokingham in Berkshire, bringing heavy rain, gusty winds and large hailstones with a diameter of more than 2.5 cm. This storm was studied in depth by scientists Keith Browning and Frank Ludlam, who produced a three-dimensional model of the supercell, explaining how large hailstones were formed within the storm.

10 July 1968 – Outbursts of thundery rain fell across southwestern England, with 127 mm recorded within 24 hours in Gloucester. The River Chew was inundated with water, sweeping bridges away in villages along its route and flooding the caves at Cheddar Gorge. Seven people died and thousands had to leave their homes.

18 July 2017 – Flash flooding devastated Coverack as thunderstorms brought more than 200 mm of rain to the coastal Cornish town within three hours. Fifty homes were flooded and residents had to be rescued by helicopter.

28 July 2005 – The strongest tornado in more than 50 years struck the city of Birmingham. Tracking nearly 12 km across the city, it tore roofs off houses, uprooted trees and injured 39 people. It was rated T5/6 on the International Tornado Intensity Scale.

JULY

In this month...

6 July – Aphelion – The Earth is at the furthest point from the Sun in its orbit.

14 July – New Moon

15 July – St Swithin's Day

29 July – Buck Moon

Look out for: Supercells
Supercells are extremely dangerous thunderstorms distinguished by a deep, rotating updraft, known as a mesocyclone. They are usually accompanied by a range of severe weather conditions, such as tornadoes, large hail, frequent lightning and damaging winds. Although they are rare in the UK, they are most likely in summer. However, they can develop at any time of year under the right conditions.

A perfect view of a supercell captured by storm chaser Chris Bell.

Observing the Weather – Noctilucent Clouds

If you look up to the night sky around midsummer, you might be lucky enough to catch a glimpse of rare noctilucent clouds. Meaning 'shining at night', noctilucent clouds, also known as polar mesospheric clouds, can only be seen a few months of the year by observers in the mid-to-high latitudes. Silvery blue in colour, they are illuminated by the Sun below the horizon, even when the Sun has already set for the observer. This is because they are much higher in the atmosphere than your regular cloud, at around 60 metres (200,000 ft) in the mesosphere, which is the next layer of the atmosphere above the stratosphere. They consist of ice crystals believed to have formed around meteoric dust from space and often resemble wispy cirrus clouds. Scientists believe the source of the ice crystals is water vapour that arrives through gaps in the troposphere or is created by chemical reactions with methane.

To spot this rare phenomenon, look northwards just after sunset on a clear summer's night. They usually become visible at the same time as the brightest stars. To get a heads-up on whether you might be able to see them on a given evening, it's worth checking social media, as cloudspotters living further east, where the Sun sets earlier, may have already seen them.

Noctilucent clouds.

Sun and Moon Times in July 2026

Location	Date	Sunrise	Sunset	Moonrise	Moonset
Belfast					
	01 Jul (Wed)	04:52	22:02	23:22	05:44
	11 Jul (Sat)	05:02	21:55	01:12	19:50
	21 Jul (Tue)	05:16	21:43	14:45	23:41
	31 Jul (Fri)	05:32	21:26	22:05	07:30
Cardiff					
	01 Jul (Wed)	04:59	21:32	22:54	05:54
	11 Jul (Sat)	05:08	21:27	01:22	19:13
	21 Jul (Tue)	05:20	21:16	14:23	23:42
	31 Jul (Fri)	05:35	21:02	21:48	07:27
Edinburgh					
	01 Jul (Wed)	04:31	22:01	23:20	05:22
	11 Jul (Sat)	04:42	21:53	00:50	19:52
	21 Jul (Tue)	04:57	21:39	14:39	23:24
	31 Jul (Fri)	05:15	21:22	21:57	07:14
London					
	01 Jul (Wed)	04:47	21:20	22:41	05:41
	11 Jul (Sat)	04:56	21:15	01:10	19:00
	21 Jul (Tue)	05:08	21:04	14:10	23:30
	31 Jul (Fri)	05:22	20:50	21:36	07:14

*all times in BST

Twilight Times in July 2026

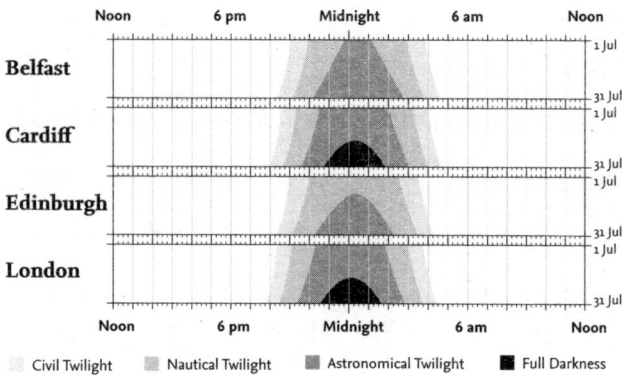

Past the solstice, our days begin to slowly shorten once again and at night, true darkness makes a return to central and southern England, Wales and parts of Ireland. Further north, nights still pass in twilight, with astronomical twilight returning to Scotland, Northern Ireland and northern England. An hour and a half of daylight is lost in the north, while in southwestern England it's closer to an hour.

Moon Phases in July 2026

Northern Hemisphere

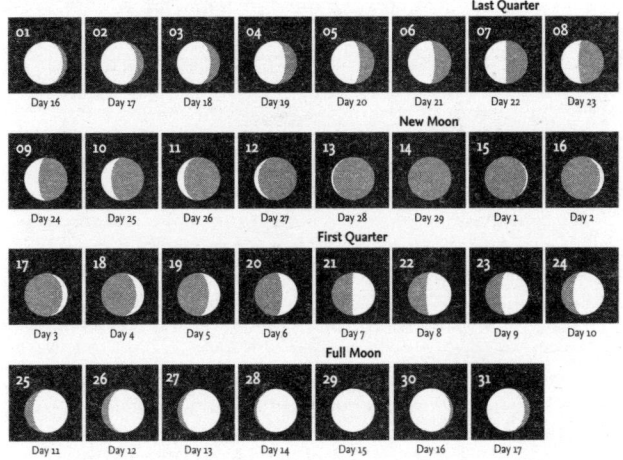

The Buck Moon
July's Full Moon marks a significant moment for male deer, who begin to regrow their antlers at this time of year after shedding them in spring. For this reason, it's known as the 'Buck Moon'. Another name for this moon marks a change in the weather instead: 'Thunder Moon', referring to the storms developing during summertime. Meanwhile, the Anglo-Saxons called it the 'Hay Moon', after the hay harvest taking place during the long, hot days of July.

The Stevenson Screen

Thermometers have come a long way since their invention. The first instruments to try and quantify temperature were thermoscopes, invented in the 1590s by Galileo. Without a scale, these devices indicated temperature change, rather than measuring it, using a container filled with bulbs of varying mass which sank or floated depending on the temperature. By 1714, Daniel Gabriel Fahrenheit produced a thermometer using mercury, and in 1724 published a scale named after himself. A couple of decades later, Anders Celsius came up with the centigrade scale, with a boiling point of 0 and freezing point of 100. After he died, the scale was inverted and the measurements we use today were born. But arguably one of the most important moments in the history of meteorological thermometers was the invention of the Stevenson screen in the nineteenth century.

Thomas Stevenson was born on 22 July 1818 in Edinburgh. A pioneering lighthouse engineer and meteorologist, Stevenson co-founded the Scottish Meteorological Society. At 24, he wrote a paper on 'Defects in Rain Gauges' and ten years later, he established a formula to predict wave heights. Perhaps his greatest contribution to the world of meteorology was the 1864 invention of a shelter to protect weather instruments from direct sunlight, while allowing air to flow freely around them. Now known as a Stevenson screen, his design quickly became popular and can be found in weather stations up and down the country.

The white, slatted screen is positioned facing northwards to ensure that no sunlight falls on the instruments when the box is opened. As well as housing a thermometer to measure air temperature, a wet-bulb thermometer is also kept inside the Stevenson screen. This is a thermometer whose tip is wrapped in a cloth soaked in water. This thermometer is cooled by evaporation, similar to the way your skin will cool when wet, and measures a lower temperature than the dry-bulb thermometer. The measurement is then used to calculate humidity.

Stevenson screens have stood the test of time, too. There has always been some concern that on calm and sunny days the screens do not ventilate properly, leading to some inaccuracies in measurements. To test this theory, scientists Professor

JULY

Giles Harrison and Dr Stephen Burt from the University of Reading recently conducted a study, comparing measurements taken in a Stevenson screen to those taken with a self-ventilating 'aspirated' thermometer. They found 'remarkably few' discrepancies and where there were errors, they were very small. Harrison and Burt concluded that, for a device more than 160 years old, the Stevenson screen continues to serve meteorologists well.

The Stevenson screen is a familiar sight at most weather stations.

Introduction

Through August, the last sultry days of late summer slip away. Typically the second-warmest month of the year, it's the home of many well-known weather records. Before the record fell in 2022, August had held the title for hottest day since records began for almost 19 years. It was in the summer of 2003 and an extreme heatwave had much of Europe in its grip. Temperatures of 25 °C or more were recorded somewhere in the UK every day from the 1st to the 18th of August, while the ten days between the 3rd and the 12th saw maximum temperatures of 30 °C or more. At the peak of the extreme heat, a large anticyclone was centred over Europe, with a southerly airflow bringing a feed of very warm continental air to the UK. On the day in question, 10 August 2003, a cold front had brought much cooler conditions to Ireland and Scotland, while thunderstorms developed over Wales and central, northern and western England. Skies stayed clear in the southeast, where temperatures soared to 38.5 °C in Faversham, Kent, and 38.1 °C in Kew, London. The former of the two temperatures received some debate about its reliability but was eventually accepted by the Met Office. It broke the previous record, also recorded in August in 1990, by more than 1 °C. The heatwave went down in history as one of the most extreme ever faced in Europe.

While August can be dry and very hot, as it was during the occasion of the 2003 heatwave, it's also likely to be unsettled and, on average, is the wettest month since February across the UK as a whole. August sometimes brings the last named storms of the season. There have been five since the Met Office began naming storms in 2015: two in 2020, two in 2023 and Storm Lilian in 2024. The latter marked the end of the most active year of storms since naming began, the Met Office never having reached the letter 'L' before. Having only started in 2015, it's not a particularly long period to draw trends from, statistically speaking, as there were certainly stormy years before naming began. However, the World Weather Attribution team, who study how climate change has influenced the intensity of extreme weather events, said rainfall during the 2023/24 storm season was 20 per cent more intense because of our changing climate.

AUGUST

August also sees plenty of convective rainfall, with some of the most notorious flash-flooding events occurring at this time of year. Many geography students who studied in the mid-noughties to the end of the 2010s will be familiar with the Boscastle floods, a devastating event that swept through a Cornish village on 16 August in 2004. A torrential 75 mm of rain fell in two hours, overwhelming the river flowing through the village and causing it to burst its banks. The extreme rainfall was part of an unusual synoptic set-up, unique to southwestern England. A line of heavy showers or thunderstorms developed on a convergence line stretching northeastwards from Bodmin Moor towards the Severn Estuary. This is known as the Brown Willy effect, named after the highest peak on Bodmin Moor.

Weather word: Dinderex (English, eighteenth century)
Where 'dinder' means thunder in old Devonshire dialect, 'dinderex' refers to a bolt of lightning, literally meaning 'thunder-axe'.

A

Weather Extremes in August

Country	Temp.	Location	Date
Maximum temperature			
England	38.5 °C	Faversham (Kent)	10 August 2003
Wales	35.2 °C	Hawarden Bridge (Flintshire)	2 August 1990
Scotland	32.9 °C	Greycrook (Scottish Borders)	9 August 2003
Northern Ireland	30.6 °C	Tandragee Ballylisk (County Armagh)	2 August 1995
Minimum temperature			
England	−2.0 °C	Kielder Castle (Northumberland)	14 August 1994
Wales	−2.8 °C	Alwen (Conwy)	29 August 1959
Scotland	−4.5 °C	Lagganlia (Inverness-shire)	21 August 1973
Northern Ireland	−1.9 °C	Katesbridge (County Down)	24 August 2014

Country	Pressure	Location	Date
Maximum pressure			
Scotland	1038.5 hPa	Stornoway (Isle of Lewis)	31 August 2021
Minimum pressure			
Republic of Ireland	966.4 hPa	Athenry (Galway)	19 August 2020

AUGUST

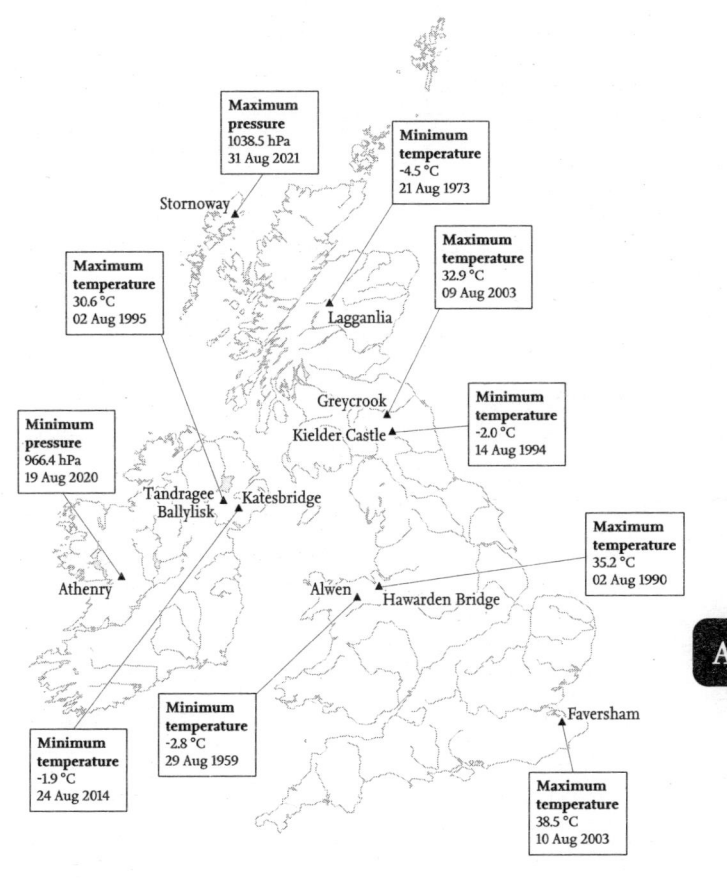

The Weather in August 2024

Observation	Location	Date
Max. temperature 34.8 °C	Cambridge NIAB (Cambridgeshire)	12 August
Min. temperature −1.2 °C	Kinbrace, Hatchery (Sutherland)	31 August
Most rainfall 154.4 mm	Honister Pass (Cumbria)	21 August
Most sunshine 14.2 hrs	Cawood (North Yorkshire)	10 August
Highest gust 74 mph (118 kph/64 kt)	Capel Curig No. 3 (Gwynedd)	22 August

Although warmer than the preceding two months, August brought a continuation of changeable conditions to end the summer, as a strong jet stream carried further low-pressure systems in from the west. The month started hot in places as the heatwave from the end of July persisted. Feeling very humid, conditions soon deteriorated with thunderstorms breaking out in the south, leading to surface-water flooding in Birmingham, Hampshire and Surrey. Temperatures gradually eased as a cold front spread eastwards, with thundery downpours in Ireland. The weather remained changeable through the following week.

Heat and humidity began to build again by the 10th as high pressure developed. The 11th was widely hot but with thunderstorms across Ireland and the far west of England and Wales. Temperatures continued to rise in the southeast on the 12th, with highs exceeding 30 °C in many places and reaching 34.8 °C at Cambridge NIAB, making this the hottest day of the year. Thunderstorms continued to move eastwards, very widespread across Scotland where lightning reportedly caused a house fire, but becoming fewer and further between

AUGUST

over England. The heat began to ease as the thunderstorms cleared.

For the rest of the month, the cool and changeable theme that characterized summer 2024 returned. The remnants of ex-hurricane Ernesto arrived in the UK on the 21st, resulting in heavy rain and strong winds. Hot on its heels, the Met Office named Storm Lilian, a deep low-pressure system that brought widespread disruption on the 23rd. Music stages were closed at the Leeds and Creamfields festivals, while people travelling to bank holiday weekend destinations were delayed at airports and on the roads. The heaviest rain fell over northern England, southern Scotland and parts of Wales while strong winds affected most places. Further depressions arrived until the 30th, when an area of high pressure brought a pleasant end to the month.

Temperatures in August were only 0.3 °C above average. Not making up for a cooler-than-average June and July, this led to summer temperatures being 0.22 °C below average, the coolest since 2015. August was also slightly wetter and duller than average but, averaged over the whole summer, rainfall and sunshine hours were close to normal.

The averages are:

Maximum temperature	19.30 °C
Minimum temperature	10.97 °C
Rainfall	93.75 mm
Sunshine	161.72 hrs
Air frost days	0.01

Historic Events

1 August 1846 – A swathe of destructive hailstorms spread over southeastern England, associated with a thundery trough. One of these storms, affecting London, was one of the most severe ever seen in the capital. Seven thousand panes of glass were smashed in the Houses of Parliament alone, with many more windows destroyed elsewhere in the city. There were also reports of a tornado, supported by the fact that a shower of frogs was observed in Fulham, lifted into the air by the powerful winds.

3 August 1829 – A catastrophic flood affected northeastern Scotland, thereafter known as the 'Muckle Spate', *muckle* meaning great in Scottish and *spate* being the word for flood. Rain, beginning the previous evening, was unrelenting through the night and day over the Cairngorms and Moray. Roads and bridges were washed away along the rivers Findhorn and Dee, eight people died and hundreds of families lost their homes.

3 August 1990 – The hottest day of the twentieth century saw mercury rise to 37.1 °C in Cheltenham, Gloucestershire. The record stood until August 2003, when temperatures reached 38.5 °C in Kent.

11 August 1999 – The last time a total solar eclipse was seen in the UK. The path of totality passed through Cornwall, where crowds flocked to witness the spectacle. Although there will be a partial solar eclipse in 2026, there won't be another total solar eclipse in the UK until 2090.

19 August 1881 – A tornado tracked across Lincolnshire over a path of 32 km, making it the longest on record in the UK. Moving from Upton to Elsham, with wind speeds up to 114 mph, it was categorized as a T3 tornado.

In this month...

12 August – New Moon

12 August – Partial solar eclipse

28 August – Partial lunar eclipse

28 August – Sturgeon Moon

Look out for: Crepuscular rays
Around sunrise or sunset you may be lucky enough to spot this optical phenomenon shining through the clouds. Crepuscular rays are shafts of sunlight peeking through the gaps in the clouds, usually seen when the Sun is just above or below a layer of cloud. In hazy conditions, the sunlight is scattered and the rays appear golden in colour. The name comes from the Latin *crepusculum*, meaning twilight, which is when they are most likely to appear.

WEATHER ALMANAC 2026

Observing the Weather – Solar Eclipse

Every 18 months or so, the chance to witness a spectacular natural phenomenon comes around. For a short space of time, in a select few locations, a solar eclipse is visible. On 12 August, that chance is coming close to the UK. Although the total eclipse path will run along a narrow path to the west of the UK, we will be treated to a partial eclipse, whereby the Moon will obscure around 90 per cent of the Sun. The last time this much of the Sun was covered in the UK was in 2015, while a total solar eclipse happened in Cornwall in 1999 and is not expected in the UK until 2090.

Since it's not safe to look directly at the Sun without proper eye protection, we need to get creative about how we watch the partial eclipse. The easiest way is to buy a pair of eclipse glasses or a handheld solar viewer with ISO 12312-2:2015 certification, which means they've been verified as a safe way to watch an eclipse. Another way is to create a pinhole projector, which can be made with common household items. Simply find two stiff pieces of card and make a small, smooth, round hole in the centre of one of them using a drawing pin. Stand with your back to the Sun, holding the card with the hole over your shoulder, so the Sun shines on and through it. Use your second piece of card as a screen. As the sun is obscured during the partial eclipse, the projection of the Sun on your second piece of card will be too. However, the projection will be flipped because the rays of sunlight intersect as they pass through the pinhole, inverting the image, like in an old-fashioned camera. You can also use a colander to project the eclipse onto a piece of card or paper.

Provided it's not too cloudy, the partial eclipse will be visible from just after 6 p.m. BST in the UK, reaching its height just after 7 p.m. and ending shortly after 8 p.m. The best place to watch will be in the southwest, where nearly 95 per cent of the Sun will be obscured.

AUGUST

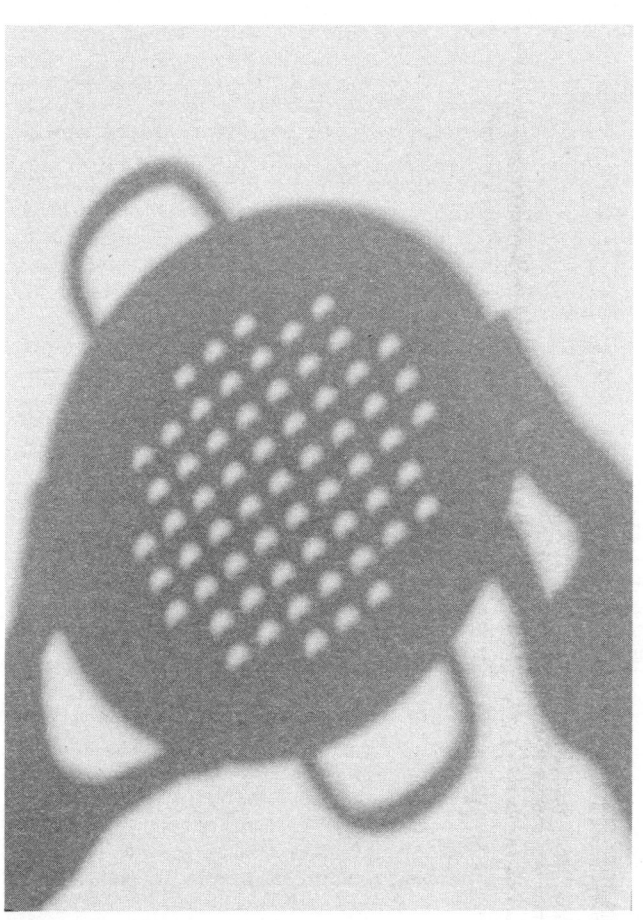

A partial solar eclipse viewed through a colander.

WEATHER ALMANAC 2026

Sun and Moon Times in August 2026

Location	Date	Sunrise	Sunset	Moonrise	Moonset
Belfast					
	01 Aug (Sat)	05:34	21:24	22:13	08:50
	11 Aug (Tue)	05:52	21:04	03:29	20:47
	21 Aug (Fri)	06:10	20:41	17:32	23:17
	31 Aug (Mon)	06:28	20:17	20:48	10:42
Cardiff					
	01 Aug (Sat)	05:36	21:00	22:00	08:43
	11 Aug (Tue)	05:51	20:42	03:37	20:21
	21 Aug (Fri)	06:07	20:22	16:53	23:33
	31 Aug (Mon)	06:23	20:01	20:45	10:23
Edinburgh					
	01 Aug (Sat)	05:16	21:20	22:03	08:36
	11 Aug (Tue)	05:36	20:58	03:07	20:44
	21 Aug (Fri)	05:55	20:34	17:34	22:52
	31 Aug (Mon)	06:15	20:09	20:33	10:33
London					
	01 Aug (Sat)	05:24	20:48	21:47	08:30
	11 Aug (Tue)	05:39	20:30	03:24	20:09
	21 Aug (Fri)	05:55	20:10	16:41	23:20
	31 Aug (Mon)	06:11	19:49	20:33	10:10

*all times in BST

AUGUST

Twilight Times in August 2026

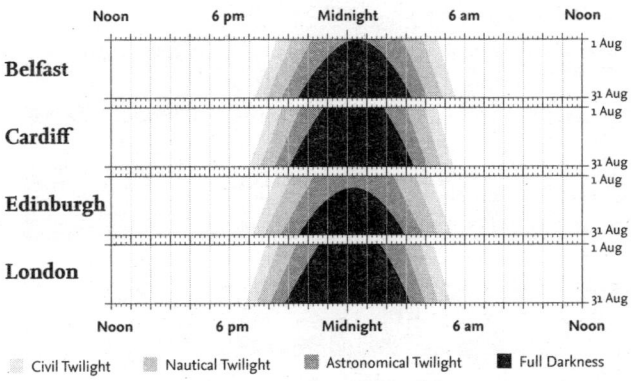

Long, sunlit days gradually slip away through August as midsummer becomes a distant memory. After several months of night-time twilight, true darkness returns to northern Scotland around mid-August as the days continue to shorten. Here, more than two hours of daylight is lost during this month, while further south the change is less pronounced, but still decreasing.

Moon Phases in August 2026

Northern Hemisphere

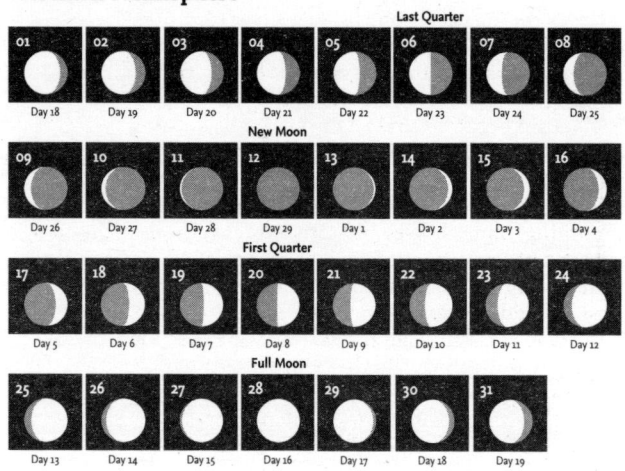

The Sturgeon Moon
Native American people named August's Full Moon after the fish they caught at this time of year. The sturgeon fish are found in the rivers, lakes and coastlines of North America, Europe and Asia. Meanwhile, Anglo-Saxons knew it as the 'Grain Moon' and other cultures referred to it as the 'Corn' or 'Green Corn Moon'.

On 28 August 2026, a partial lunar eclipse will give the Full Moon a reddish tinge as it falls into Earth's shadow.

AUGUST

The Great Flood of 1912

As the UK's coldest, wettest and dullest summer on record drew to a close, an unusual synoptic situation began to brew over France and the Netherlands. A small depression had developed and began to track northwards over the Strait of Dover, rapidly deepening as it did so, towards the unsuspecting county of Norfolk. Rain began around 3 a.m. on Monday, 26 August 1912. By the time the residents of Norwich awoke to a cool and wet day, 26 mm of rain had already fallen, but little did they know how much more was to come. Business continued as normal through the morning; trams, horses and carts trundled through the city streets and pedestrians attempted to get on with their business, despite torrential rain rendering umbrellas and raincoats useless. By midday, train services had ground to a halt as the rain continued to fall 'as though the windows of Heaven had opened', according to one record. Winds whipped up to gale force and the city came to a standstill as streets turned to rivers and still the rain did not abate. By the following morning, more than 164 mm of rain had fallen in Norwich, while over 200 mm had been recorded to the east of the city. Although the rain began to ease, rivers continued to rise. Floodwaters attempting to flow out into the sea were met by abnormally high tides at Great Yarmouth, effectively blocking their exit. The River Wensum in the centre of Norwich rose to five metres above the mean high level, leaving many homes underwater. Some of the worst affected were those living in slums to the north of the river. Sewers burst and roads cracked and subsided. Rescues were carried out by boat and seven schools opened their doors as emergency shelters, as well as several clubs and charities. To top it all, the city lost all light and power that evening, as floodwaters made their way into the electricity station. Norwich was completely cut off, not only by road and rail, as 52 bridges across Norfolk had been washed away, but by telephone and telegraph too. Around 15,000 people were affected by the devastating floods, four people lost their lives and farmers across Norfolk lost nearly all their corn crop. Over the following days, floodwaters began to recede and the county started to recover from the extreme weather. Glimmers of hope include stories of heroic rescues, emergency food parcels including chocolate bars for more than 1,000 affected households, and twelve babies safely born in shelters. Scientists studying the event years later suggested the torrential rains may have been influenced by local airflow convergence and warm water around the Thames Estuary.

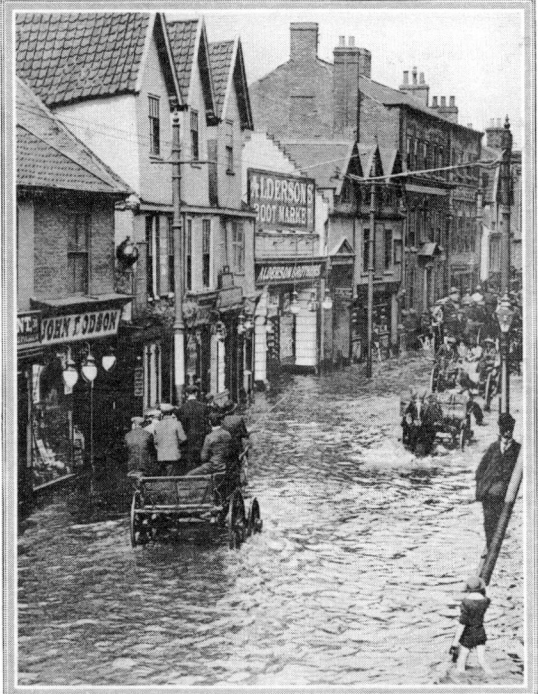

'Norwich transformed into Venice' during the floods of August 1912.

Introduction

The last days of summer melt into the first days of autumn as hedgerows hang with juicy blackberries and the Sun shines on golden fields of crops ready to be harvested. Although it often feels like summer hasn't left, the first day of September marks the beginning of meteorological autumn, with astronomical autumn not far behind. Typically more settled than the months either side of it, September can bring a last hurrah and chance to enjoy a few more days of warmth and sunshine before autumn well and truly sets in. Every now and then, September will even be home to the hottest day of the year. This happened in 2023, with seven consecutive days reaching 30 °C somewhere in the UK from the 4th to the 10th, the heat culminating in a peak of 33.5 °C on the last day of the heatwave. It was only the fifth time since records began that the warmest day of the year was in September and it helped the month secure the joint top spot of warmest September on record, tied with 2006.

More often though, September sees the start of the gradual decline in temperatures. Nights begin to draw in and you may even find yourself reaching for a coat in the chilly morning air, only to cast it off by lunchtime. Gardeners are on the lookout for the first grass frost of the season, the chance of which begins to creep up at this time of year, although many won't see a frost until later in the autumn. The coldest September on record, in 1952, was well remembered for its frosty nights, with a widespread ground frost on the 7th and a daytime high of only 9 °C in Dunstable, Bedfordshire, the lowest to be recorded in the first ten days of September at the time. Another such September was that of 1986: cold, anticyclonic and the frostiest on record thanks to frequent clear skies at night. September 2024 also had a remarkably early first frost of the season (see page 150).

Towards the end of September, astronomical autumn begins with the equinox. Weather lore suggests that around this time there's also an upswing in storms and these are sometimes referred to as 'equinoctial gales'. Thought to have developed from the observations of mariners, these peaks in stormy weather are more fiction than fact, as they're not supported by climatological evidence. However, September does mark the start of a new storm season. Each year since 2015, the UK

SEPTEMBER

Met Office, in collaboration with Ireland's Met Éireann and the Dutch meteorological service, KNMI, publishes a list of names to be used in the upcoming storm season to help raise awareness of these hazards and their potential impacts.

Weather word: Gloaming, also Gloamin (Scottish, nineteenth century)
Evening twilight or dusk, as the nights begin to draw in.

Weather Extremes in September

Country	Temp.	Location	Date
Maximum temperature			
England	35.6 °C	Bawtry, Hesley Hall (South Yorkshire)	2 September 1906
Wales	32.3 °C	Hawarden Bridge (Flintshire)	1 September 1906
Scotland	32.2 °C	Gordon Castle (Moray)	1 September 1906
Northern Ireland	28.0 °C	Castlederg (County Tyrone)	8 September 2023
Minimum temperature			
England	−5.6 °C	Santon Downham (Suffolk) Grendon Underwood (Buckinghamshire)	30 September 1969
Wales	−5.5 °C	St Harmon (Powys)	19 September 1986
Scotland	−6.7 °C	Dalwhinnie (Inverness-shire)	26 September 1942
Northern Ireland	−3.7 °C	Katesbridge (County Down)	27 September 2020

Country	Pressure	Location	Date
Maximum pressure			
Northern Ireland	1042.0 hPa	Ballykelly (County Londonderry)	11 September 2009
Minimum pressure			
Republic of Ireland	957.3 hPa	Claremorris (County Mayo)	21 September 1953

SEPTEMBER

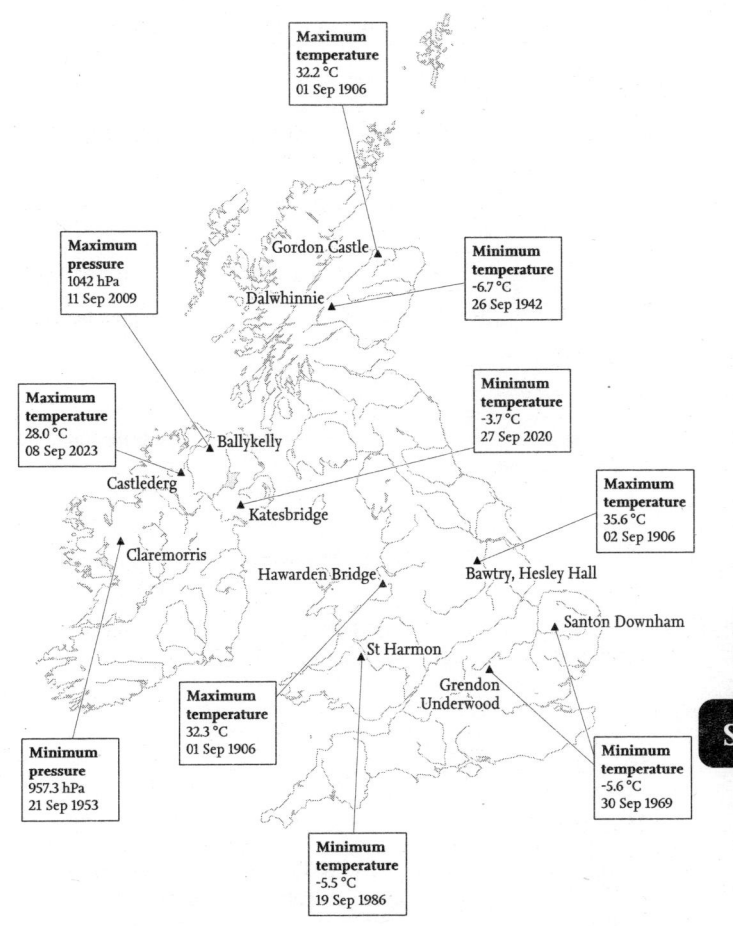

The Weather in September 2024

Observation	Location	Date
Max. temperature 30.1 °C	Cambridge University Botanic Garden (Cambridgeshire)	1 September
Min. temperature −3.0 °C	Tyndrum No. 3 (Perthshire) and Braemar No. 2 (Aberdeenshire)	25 September
Most rainfall 119.9 mm	White Barrow (Devon)	29 September
Most sunshine 12.7 hrs	Stornoway Airport (Western Isles)	5 September
Highest gust 69 mph (110 kph/60 kt)	Berry Head (Devon)	29 September

Bucking the trend of being one of the UK's typically more settled autumn months, September 2024 will be remembered for its widespread flooding across central England. Two of the worst-affected counties were Bedfordshire and Oxfordshire, which not only saw their wettest September, but also their wettest calendar month on record.

The month began as it meant to go on: unsettled. It also started warm, recording its highest temperature on the first day of the month, 30.1 °C in Cambridge. The weather was characterized by widespread showers and sporadic thunderstorms, driven by a warm and humid southerly airflow. However, northerly winds developing around the 11th brought the first taste of cooler weather. Arctic air moved in and temperatures plunged to between 3 and 6 °C below average for the time of year. Some areas saw their earliest first frosts in a number of years. On the morning of the 14th, Santon Downham, a Suffolk weather station notorious for recording extremes due to its light, sandy soil, recorded a minimum temperature of -0.3 °C, the earliest September air frost in East Anglia since 1986. It wasn't until high pressure built mid-month that widespread dry, settled and warm conditions prevailed. Much of the UK basked in autumnal sunshine once early morning mist and fog cleared. Under clear skies, many were able to catch a glimpse of the aurora borealis, as well as the Harvest supermoon. Unsettled weather returned for the last ten days of September. Several tornadoes were reported across southern England: in Hampshire on the 20th and 26th, and near Luton in Bedfordshire on the 22nd. From this date began a week of severe flooding, with Woburn in Bedfordshire measuring a staggering 120.4 mm of rain in just 30 hours. It's no surprise that this station was the wettest place in the UK in September.

Averaged across the whole month, temperatures were −0.3 °C below average. Southern England was exceptionally wet, recording 244 per cent of average rainfall. However, with Scotland receiving only 63 per cent of average rainfall, the overall rainfall for the UK was 125 per cent of average. Similarly, sunshine was above average in the north and below average in the south, resulting in close-to-average totals overall.

The averages are:

Maximum temperature	16.85 °C
Minimum temperature	9.04 °C
Rainfall	90.90 mm
Sunshine	127.51 hrs
Air frost days	0.12

WEATHER ALMANAC 2026

Historic Events

2 September 1906 – A heatwave beginning in late August peaked on the second day of meteorological autumn. In Bawtry, South Yorkshire, a high of 35.6 °C was reached, making it the hottest September day on record.

2 September 1974 – Gale-force winds whipped through the English Channel, bringing rough seas and a storm surge. The severe weather wrecked Prime Minister Edward Heath's yacht, the *Morning Cloud*, which was travelling from Burnham-on-Crouch to Cowes, and two crew members lost their lives. It was thereafter known as the 'Morning Cloud Storm'.

15 September 2009 – A narrow band of intense rainfall stretching from Suffolk to Dorset brought torrential downpours to parts of southern England. In some places, 40–60 mm of rain fell, with localized flooding disrupting travel and sports fixtures, including the football match between Queens Park Rangers and Crystal Palace.

19 September 2003 – A burst of exceptionally heavy rain caused a landslide at Pollatomish, County Mayo, Ireland. A section of Dooncarton mountain, which overlooks the village, collapsed, blocking roads, damaging homes and destroying bridges, leaving hundreds of locals stranded without power or water.

22 September 1935 – A powerful and damaging hailstorm tracked 335 km across the UK, bringing a swathe of destruction from Newport in South Wales to Mundesley in north Norfolk.

Hail as large as a fist was reported near Northampton, where the storm was known as the 'Great Northamptonshire Hailstorm'. With extensive damage to glass, roofs and telegraph wires, it remains the longest hailstorm ever recorded in the UK.

In this month...

1 September – Meteorological autumn begins

11 September – New Moon

13 September – World Cloud Appreciation Day

23 September – Astronomical autumn begins

23 September – Autumn equinox

26 September – Full Corn & Harvest Moon

Look out for: Extratropical cyclones
The technical name for what we know as storms or depressions in the UK, a 'cyclone' describes any circulation of air around a low-pressure centre, while 'extratropical' refers to its location in the mid-to-high latitudes, where the UK is found. In the northern hemisphere, winds circulate around the centre of the cyclone in an anticlockwise direction, while the opposite is true in the southern hemisphere. The more the pressure drops in the centre of the extratropical cyclone, the more intense it becomes. If pressure drops by more than 24 millibars in 24 hours, this is known as 'explosive cyclogenesis' or a 'weather bomb'.

Observing the Weather – The Beaufort Scale

Generally, an anemometer is needed to accurately measure the wind speed and the first image to come to mind is the classic device with three cups rotated by the blowing wind. As the measurement is taken, the speed of the wind is calculated from the rate of rotation of the cups. However, in 1805, Irish hydrographer and naval officer Francis Beaufort invented a scale that allows weather enthusiasts and meteorologists alike to estimate wind speeds. He related wind speeds to a range of observable conditions, which could be made on land or sea. On land, they refer to the movement of smoke rising from a chimney, the movement of leaves and branches, and at the extreme end, damage to buildings. Meanwhile, the scale used at sea describes how the waves change as wind speeds increase. The Beaufort scale for both land and sea are included on pages 249–251.

Its first official use was on HMS *Beagle*, captained by Robert FitzRoy, who went on to set up the first Meteorological Office in the UK.

The Beaufort scale is still used today by sailors and meteorologists in marine forecasts.

Rear-Admiral Sir Francis Beaufort.

Sun and Moon Times in September 2026

Location	Date	Sunrise	Sunset	Moonrise	Moonset
Belfast					
	01 Sep (Tue)	06:30	20:15	21:00	12:10
	11 Sep (Fri)	06:49	19:50	07:10	19:39
	21 Sep (Mon)	07:07	19:25	17:57	00:11
	30 Sep (Wed)	07:23	19:02	19:49	12:56
Cardiff					
	01 Sep (Tue)	06:25	19:59	21:02	11:46
	11 Sep (Fri)	06:41	19:36	07:00	19:28
	21 Sep (Mon)	06:57	19:13	17:28	00:22
	30 Sep (Wed)	07:11	18:52	19:59	12:24
Edinburgh					
	01 Sep (Tue)	06:17	20:07	20:43	12:04
	11 Sep (Fri)	06:36	19:41	06:58	19:27
	21 Sep (Mon)	06:55	19:14	17:54	01:10
	30 Sep (Wed)	07:13	18:50	19:27	12:55
London					
	01 Sep (Tue)	06:13	19:46	20:50	11:34
	11 Sep (Fri)	06:28	19:24	06:47	19:16
	21 Sep (Mon)	06:44	19:01	17:15	00:09
	30 Sep (Wed)	06:59	18:40	19:47	12:11

*all times in BST

WEATHER ALMANAC 2026

Twilight Times in September 2026

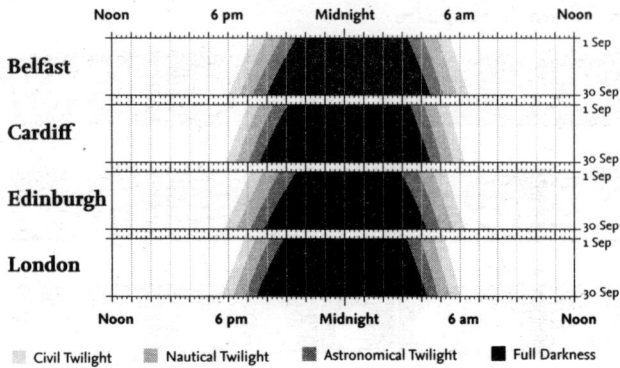

On 23 September, the autumn equinox marks the moment the Sun appears to cross the Equator and night and day are supposedly equal. However, because of the size of the Sun and the way its light reaches us, day and night are truly equal a couple of days after the equinox. This is called the equilux and from hereafter, the nights become longer than the days. Throughout the whole of September around two hours of daylight are lost, depending on latitude.

Moon Phases in September 2026

Northern Hemisphere

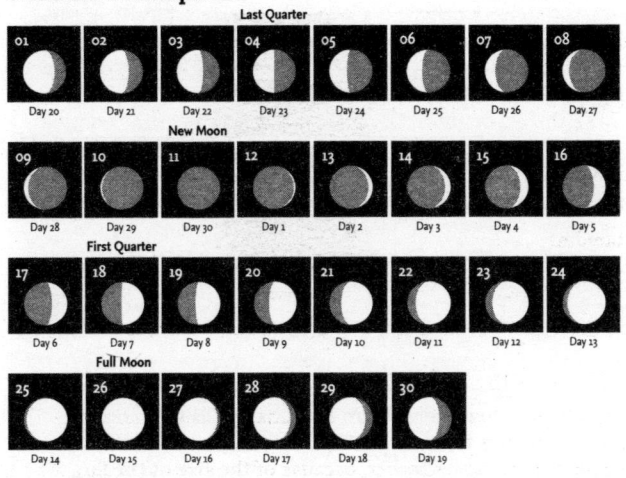

The Harvest & Full Corn Moon
As the closest Full Moon to the autumn equinox, September's Moon is known as the 'Harvest Moon', marking an important time of year for farmers. As the equinox is in September, the Harvest Moon nearly always is too. However, on some occasions the Harvest Moon falls in October, and in that case the September Full Moon is simply known as the 'Full Corn Moon'.

Hurricanes

While in the UK our storm season is only just beginning in September, it's a different story across, or technically in, the pond. The Atlantic hurricane season begins in June, bringing the threat of tropical cyclones to those living in the Caribbean, the Gulf of Mexico and the east coast of the United States. These storms begin life over the warm waters of the tropical Atlantic Ocean, as winds begin to rotate around a central area of low pressure. At this stage it's known as a tropical cyclone, but when wind speeds reach 74 mph it becomes a hurricane. Typically, 14 Atlantic storms are named by the US National Hurricane Center each year. On average, seven of these develop into hurricanes and three become major hurricanes of category 3 or higher, when wind speeds reach more than 111 mph. The peak of the Atlantic hurricane season is 10 September, with the first major hurricane typically named in the first few days of the month.

One of the key conditions for the formation of a hurricane is sea temperatures of at least 27 °C, which means we never get them back home as our seas aren't warm enough, and once a hurricane travels out of the tropics, it loses its status as a tropical cyclone. However, its remnants can get wrapped up into our extratropical cyclones and are known as ex-hurricanes. One of the most destructive storms to affect the UK and Ireland was ex-hurricane Debbie, which wrought havoc across Ireland in September 1961. Hurricane Debbie was named on 6 September, developing unusually far east, just off the Cape Verde Islands. Tracking northwestwards, it reached category 3 status on 11 September, whereby it pivoted and began moving towards Ireland. As it made its way over the cooler water of the North Atlantic it lost some of its energy. Although no longer a hurricane, it was still a very powerful storm with many similar characteristics as it passed close to the west coast of Ireland on Saturday, 16 September. Ex-hurricane Debbie brought sustained wind speeds of 76 mph and gusts of up to 113 mph to Malin Head, the most northerly point of mainland Ireland. The hurricane-force winds claimed the lives of 18 people across Northern Ireland and the Republic of Ireland, who were mostly killed by falling trees. The storm caused widespread

SEPTEMBER

damage, especially to northern and western areas where few buildings escaped damage and hundreds of people were treated in hospital for injuries caused by flying debris. Ex-hurricane Debbie went down as the most severe storm to affect Ireland in modern memory.

Storm damage from ex-hurricane Debbie in Ireland.

Introduction

Despite its reputation for increasingly cold, damp weather, October makes up for the occasional gloom with beautiful autumn imagery. Fat, shiny conkers and crisp, round apples fall from the trees, whose leaves are transforming from their summer green to beautiful shades of gold, russet and deep crimson. Leaves change colour as their levels of the green pigment, chlorophyll, decrease and other pigments begin to take over. The weather can influence this natural process, determining what colours our leaves might turn. A warm, dry and sunny autumn will cloak the trees in leaves of red, while sub-zero nights turn them yellow. However, strong winds and heavy rain, which become more likely as October progresses, can strip trees of their leaves before we've had a chance to enjoy them.

It's at this time of year that the jet stream begins to reinvigorate. The growing temperature contrast between the North Pole and the Equator adds extra energy to it, making the jet stream stronger and more zonal (flowing in a straight(ish) line from west to east) rather than meridional (which means meandering from north to south). This strengthening jet helps deepen areas of low pressure and push them towards the UK, bringing an upswing in wet and windy weather.

Another change marked in October is the end of the land-based convection season. This is the period of time between late winter/early spring and mid-autumn, when showers or thunderstorms can be generated by the heating of air close to the Earth's surface, which causes it to rise, cool and condense. From around mid-October, the heat reaching us from the Sun is no longer strong enough to trigger this process, so we say goodbye to land-based showers, thunderstorms and fair-weather clouds until February next year. However, thunderstorms are still possible, provided that one of the key ingredients, instability, can be generated another way. One of these is the relatively warm waters around the UK. While the land is cooling steadily through autumn, our surrounding seas hold on to their warmth for longer and are able to bring instability to an air mass. Air just above the sea is warmed, and rises in the same way that this happens over land in summer, so showers are generated over the water instead. In summer, showers are more likely to form during the day when the Sun is out to provide heat, whereas in winter the ocean provides the warmth, so showers can develop at any time. On one October night in 2008, a small town in Devon was hit by a dramatic hailstorm, which persisted for two hours and buried Ottery St Mary in 30 cm of hailstones. This exceptional storm is thought to have been caused by the relatively high sea surface temperatures of late October, around 14 °C, and uplift provided by local topography. Alongside the incredibly long-lasting hailstorm, around 200 mm of rain fell over Ottery St Mary in the early hours of the 30th. Hail blocked waterways and ditches, forming ice floes and leading to the River Otter bursting its banks. The town became cut off and more than 100 people had to be evacuated from their homes. A local newspaper reported that the storm and its aftermath cost the town £1 million.

Weather word: Drookit, also Drook (Scottish)
Rain that drenches, soaks or wets through to the skin.

Weather Extremes in October

Country	Temp.	Location	Date
Maximum temperature			
England	29.9 °C	Gravesend (Kent)	1 October 2011
Wales	28.2 °C	Hawarden Airport (Flintshire)	1 October 2011
Scotland	27.4 °C	Tillypronie (Aberdeenshire)	3 October 1908
Northern Ireland	24.1 °C	Strabane (County Tyrone)	10 October 1969
Minimum temperature			
England	−10.6 °C	Wark (Northumberland)	17 October 1993
Wales	−9.4 °C	Rhayader, Penvalley (Powys)	26 October 1931
Scotland	−11.7 °C	Dalwhinne (Inverness-shire)	28 October 1948
Northern Ireland	−7.2 °C	Lough Navar Forest (County Fermanagh)	18 October 1993

Country	Pressure	Location	Date
Maximum pressure			
Scotland	1045.6 hPa	Dyce (Aberdeenshire)	31 October 1956
Minimum pressure			
Scotland	946.8 hPa	Cawdor Castle (Nairnshire)	14 October 1891

OCTOBER

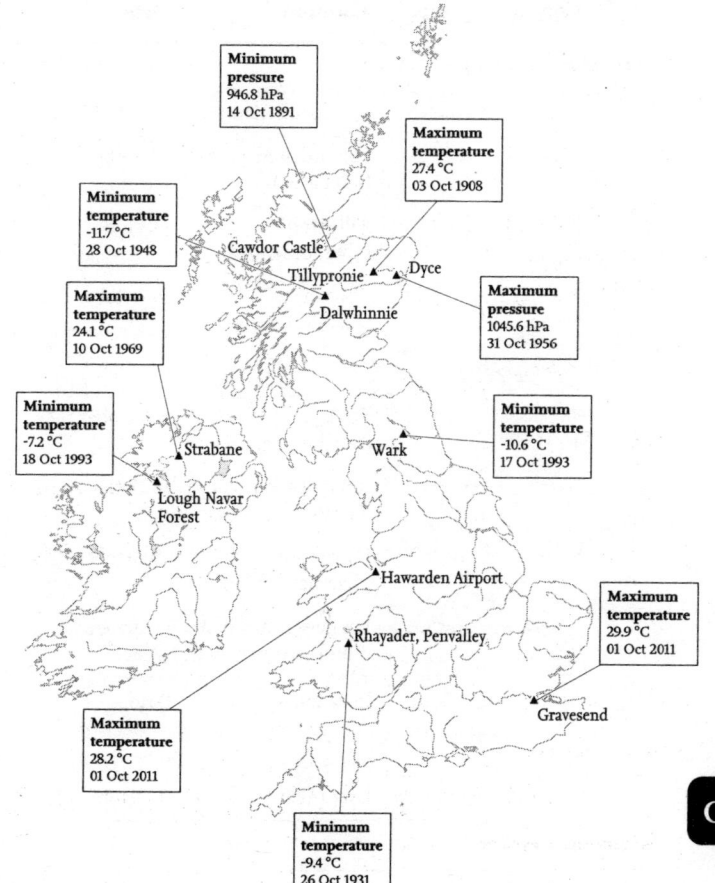

The Weather in October 2024

Observation	Location	Date
Max. temperature 22.5 °C	St James's Park (Greater London)	16 October
Min. temperature −4.1 °C	Braemar No. 2 (Aberdeenshire) and Aboyne No. 2 (Aberdeenshire)	3 October 14 October
Most rainfall 74.2 mm	Ulpha, Duddon Water Works (Cumbria)	27 October
Most sunshine 10.6 hrs	Exeter Airport No. 2 (Devon)	3 October
Highest gust 82 mph (131 kph/71 kt)	Aberdaron (Gwynedd)	20 October

October 2024 was mild and fairly mixed. The month saw a succession of low-pressure systems moving eastwards across the UK, interspersed with spells of drier, more settled weather brought by brief ridges of high pressure.

It began wet and unsettled across central and eastern England, exacerbating the previous month's flooding. Conditions turned drier as high pressure built briefly over Scotland, but the next depression soon moved in from the Atlantic on the 6th, bringing further wet and windy weather. A few days later, ex-Hurricane Kirk arrived. Tracking to the south, the UK avoided the worst impacts but was still rather wet. As it cleared it brought a change in wind direction and the resultant northerly winds led to a spell of colder conditions around the 10th. Skies also cleared and a well-timed coronal mass ejection allowed people as far south as Oxford to catch a glimpse of the aurora borealis. The cool, settled weather was short-lived and fronts soon began to move back in, bringing outbreaks of heavy and persistent rain around mid-month. On 18 October, the Irish Meteorological Service, Met Éireann, named the first storm of the 2024/25 season. Storm Ashley arrived on the 20th, bringing strong winds and heavy rain to the north and west. Gusts of between 60 and 70 mph were recorded widely across Scotland, Ireland, northwestern England and Wales. Exposed areas and high ground saw gusts exceeding 80 mph, reaching more than 100 mph over the highest peaks. The rest of the month passed in a similar vein, with various fronts moving eastwards across the UK and occasional spells of quieter, more settled conditions. Towards the end of the month, high pressure began to build from the south, bringing a dry end to a variable October.

With much of the month dominated by Atlantic conditions, it was almost always mild, with temperatures 0.7 °C above the long-term average. Despite its changeable weather, October was also drier than usual, with the UK receiving only 84 per cent of its average rainfall. However, this is averaged across the whole country, and at a local level some parts of southern, central and northeastern England still recorded a wetter-than-average month. Sunshine was close to average.

The averages are:

Maximum temperature	13.08 °C
Minimum temperature	6.42 °C
Rainfall	122.52 mm
Sunshine	91.85 hrs
Air frost days	1.64

Historic Events

14 October 1881 – A rapidly deepening depression brought gale-force winds and rough seas to the coasts of southeastern Scotland. A total of 189 fishermen, who had headed out to sea earlier in the day, never came home, leading to the Eyemouth fishing disaster, also known locally as 'Black Friday'.

15 October 1962 – Having included forecast temperatures in degrees Celsius for over a year previously, British weather forecasts made the official switch from the Fahrenheit scale to Celsius.

21 October 1638 – One of the first recorded sightings of 'ball lightning', a phenomenon characterized by a bright, roughly spherical ball varying in colour from red to yellow, white or blue. Eyewitness accounts describe it bursting through the window of a church during the 'Great Thunderstorm of Widecombe-in-the-Moor' on Dartmoor. Several members of the congregation were killed and many were injured.

23 October 1091 – The earliest-known tornado hit central London. The church of St Mary-le-Bow suffered extensive damage and around 600 mostly wooden houses were destroyed. It was equivalent to T8 on the International Tornado Intensity Scale and is one of the most intense on record in the UK.

31 October 1950 – Thick fog caused a plane crash at London Airport (now known as Heathrow). As the Vickers Viking plane, carrying 30 passengers and crew, attempted to land during low visibility of only 37 metres, it struck the runway unexpectedly, damaging the two propellers and tearing off the starboard wing. The weather hampered rescue attempts as it took more than 15 minutes for aid to find the crash site in the heavy fog. There were only two survivors.

In this month...

10 October – New Moon

25 October – British Summer Time ends

26 October – Hunter's Moon

Look out for: Line convection
While land-based convection ceases through late autumn and winter, other types of convection can bring heavy rainfall and squally winds. Line convection describes a band of intense rain and gales which develops along a zone of rapidly rising air within a cold front. It can be seen on a weather radar as a thin, brightly coloured line within a band of rain – sometimes broken up into lots of little lines tilted slightly forward.

Observing the Weather – Making a Rain Gauge

As the wetter half of the year begins, it's the perfect time to make a rain gauge. One of the simplest weather experiments to conduct, it requires only a few household items.

You will need:

- Clear plastic water bottle – a large 2-litre bottle is ideal
- Scissors
- Sticky tape
- Ruler or tape measure
- Marker pen
- Water

First, using scissors, cut your water bottle horizontally about one-third down from the top. Because most plastic water bottles don't have a flat bottom, measure out 100 mm of water and pour it into your bottle. Resting the bottle on a level surface, use your marker pen to draw a continuous line around the bottle marking the water level – this will be the bottom of your scale. Next, use a ruler held against the bottle to mark measurements on your bottle, creating a scale. You may prefer to mark this on some sticky tape and then stick it on. Increments of 0.5 cm work well, but if you have a finer pen and a steady hand feel free to be even more precise. Once you've finished your scale, take the top third of your bottle, flip it upside down and fit it into the bottom part of your bottle, creating a funnel for rain to fall into. Fix it in place with some sticky tape. Now all that's left is to place your rain gauge outside. Choose a location that isn't too sheltered or close to tall structures. To keep the rain gauge upright, you may either dig a small hole, surround it with large stones or bricks, or place it in a bucket of sand. Take your measurements at the same time every day and try comparing it with a local weather station to see how closely yours match.

OCTOBER

A homemade rain gauge measuring 40 mm of rain.

WEATHER ALMANAC 2026

Sun and Moon Times in October 2026

Location	Date	Sunrise	Sunset	Moonrise	Moonset
Belfast					
	01 Oct (Thu)	07:25	18:59	20:28	14:24
	11 Oct (Sun)	07:44	18:35	08:57	18:15
	21 Oct (Wed)	08:04	18:11	16:39	01:48
	31 Oct (Sat)	07:24	16:49	21:06	13:58
Cardiff					
	01 Oct (Thu)	07:13	18:50	20:43	13:47
	11 Oct (Sun)	07:29	19:28	08:36	18:14
	21 Oct (Wed)	07:46	18:07	16:21	01:46
	31 Oct (Sat)	07:04	16:47	21:13	13:26
Edinburgh					
	01 Oct (Thu)	07:15	18:48	20:03	14:26
	11 Oct (Sun)	07:35	18:22	08:50	17:58
	21 Oct (Wed)	07:56	17:57	16:31	01:32
	31 Oct (Sat)	07:17	16:34	20:44	13:57
London					
	01 Oct (Thu)	07:01	18:38	20:30	13:34
	11 Oct (Sun)	07:17	18:15	08:23	18:02
	21 Oct (Wed)	07:34	15:54	16:09	01:33
	31 Oct (Sat)	06:52	16:35	21:00	13:14

*times change to GMT on 25 October

Twilight Times in October 2026

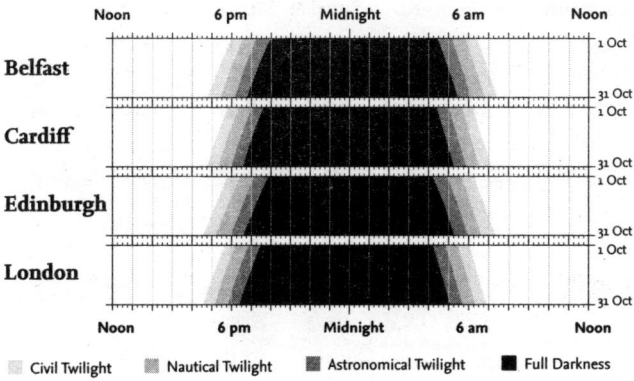

Darkness opens its arms in an embrace as more than two hours of daylight are lost during October. This change is particularly pronounced in northern Scotland, where the days shorten from around 11.5 hours at the beginning of the month to just over nine hours by the end of October. On the 25th, the clocks 'fall back', bringing lighter mornings but darker evenings, as the Sun sets an hour earlier.

Moon Phases in October 2026

Northern Hemisphere

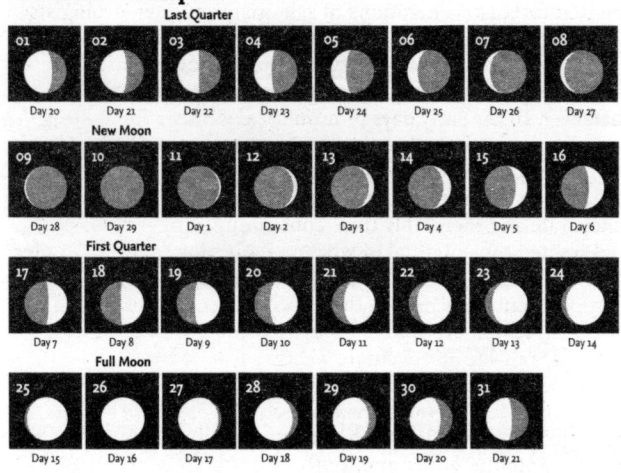

The Hunter's Moon
Historically in the northern hemisphere, October was a time to prepare for the approaching winter, by hunting, gathering and preserving food under the light of the full 'Hunter's Moon'. Once providing good hiding places for hunted animals, fields had been harvested and cleared by this time of year, allowing hunters to easily catch their prey.

OCTOBER

Artificial Intelligence and Weather

Since the first numerical weather prediction model churned out the first forecast, generations of scientists and meteorologists have continued to innovate, inspired by the possibility of gleaning the future. Using new techniques and equations, they have created weather forecasts that are more accurate than ever. In the early days of numerical weather forecasting, complex mathematical equations were carried out by hand. In 1922, English mathematician and meteorologist Lewis Fry Richardson published the first forecast. Almost completely inaccurate and incredibly time-consuming, it nevertheless highlighted the potential of what could be done. Three decades later, both American and British meteorologists succeeded in using computers to resolve their complex equations and in the following years were able to hone their models to produce a relatively reliable and accurate weather forecast.

Nowadays, immensely powerful supercomputers carry out quadrillions of equations per second. They power large-scale global models, long-range forecasts looking weeks, if not months, into the future and high-resolution models capable of capturing small-scale convective systems, such as showers and thunderstorms.

As humanity enters a new age of artificial intelligence (AI), imaginative scientists are already investigating how machine learning can benefit our weather forecasts. The concept involves detecting patterns from past weather data to attempt to forecast the future, rather than using physical equations to extrapolate from current weather observations . The main advantage of these AI weather models is that they are much quicker and cheaper to run, delivering forecasts in a fraction of the time it takes for the traditional models, particularly in areas with fewer or less frequent observations.

In October 2023, the Met Office and the Alan Turing Institute announced a partnership programme, AI4NWP, to harness the power of AI in weather forecasting. Together, they created the FastNet machine-learning weather model – named after an area in the shipping forecast, which was established by the Met Office's founder, Robert FitzRoy. The model shows plenty of promise, with results comparable to the Met Office's

current numerical weather prediction system. The AI4NWP team hopes to build on this initial success and introduce the model into operational forecasts in the years to come.
The Met Office is not the only one developing these state-of-the-art models. Many other forecasting and technology companies have already created or started the process of making their own AI weather models. In early 2025, The European Centre for Medium-Range Weather Forecasts (ECMWF) launched its Artificial Intelligence Forecasting System (AIFS), while technology companies Google and Huawei are also running AI weather predictions.

The Met Office Cray XC40 supercomputing system.

Introduction

A mixture of cold, foggy mornings and rain lashing the windowpanes: November is here. One of the more unsettled months of the year, a typical November often sees a string of depressions moving in from the Atlantic, interspersed with blasts of northerly air and occasional settled intervals as lows clear eastwards. As well as frequent rain, the month is almost synonymous with fog. Sometimes just an early-morning blanket over the valley floors and other times a thick, sun-obscuring entity that gives the day an almost dreamlike quality. Fog is made of millions of tiny water droplets suspended in the air. Each water droplet is condensed around a miniscule particle of dust or air pollution, known as cloud condensation nuclei. Fog has to be thick enough to earn its name, reducing visibility to one kilometre or less. Higher than that and it's known as mist instead. Although not necessarily the foggiest month of the year – December often vies for that crown – there's a good reason many of us associate November with this weather type: Bonfire Night. Also known as Guy Fawkes Night, it's celebrated on 5 November and often for a week before and after. Fireworks fill the night skies with noise, colour and light, and bonfires, often topped with an effigy of Guy Fawkes, are burned to mark the downfall of the 1605 Gunpowder Plot. The event results in large amounts of smoke being released into our lower atmosphere, meaning there are now millions more cloud condensation nuclei for fog to form on. However, a few more conditions need to fall into place in order for fog to develop – we need the right weather. Clear skies and light winds are perfect, allowing temperatures to fall and water vapour in the air to condense onto the smoke particles to form a thick blanket of fog.

   As well as mist and fog, November is likely to bring the first widespread frosts and the increasing potential of snow, especially to northern areas. The coldest November on record was in 1919, largely thanks to a bitter spell of weather mid-month. From the 11th to the 16th, weather reports describe the period as 'unprecedentedly cold' with widespread ground and hoar frosts across the UK and frequent snow, sleet and hail. On the 12th, a severe snowstorm brought 43 cm of snow to

Balmoral, Aberdeenshire, which lay on the ground for the rest of the month. Overall, temperatures were more than 3 °C below average in November 1919. However, the winter that followed was remarkably dull and mild. This example lives up to the English weather lore rhyme: 'If there's ice in November that will bear a duck, there will be nothing after but sludge and muck.' There are many factors that can influence our winters here in the UK (see page 6–7) and one of these is the El Niño Southern Oscillation. In its cold phase, La Niña, some scientists believe our winters will begin cold, then turn milder, so perhaps that was how the lore came about.

Weather word: Bullet staines (Scottish)
Hailstones, falling like bullets.

Weather Extremes in November

Country	Temp.	Location	Date
Maximum temperature			
England	21.1 °C	Chelmsford (Essex) Clacton (Essex) Cambridge (Cambridgeshire) Mildenhall (Suffolk)	5 November 1938
Wales	22.4 °C	Trawsgoed (Ceredigion)	1 November 2015
Scotland	20.6 °C	Edinburgh Royal Botanic Garden Liberton (Edinburgh)	4 November 1946
Northern Ireland	18.5 °C	Murlough (County Down)	3 November 1979 1 November 2007 and 10 November 2015
Minimum temperature			
England	−16.1 °C	Scaleby (Cumbria)	30 November 1912
Wales	−18.0 °C	Llysdinam (Powys)	28 November 2010
Scotland	−23.3 °C	Braemar (Aberdeenshire)	14 November 1919
Northern Ireland	−12.2 °C	Lisburn (County Antrim)	15 November 1919

Country	Pressure	Location	Date
Maximum pressure			
Scotland	1046.7 hPa	Aviemore (Inverness-shire)	10 November 1999
Minimum pressure			
Republic of Ireland	931.2 hPa	Limerick (County Limerick)	11 November 1877

NOVEMBER

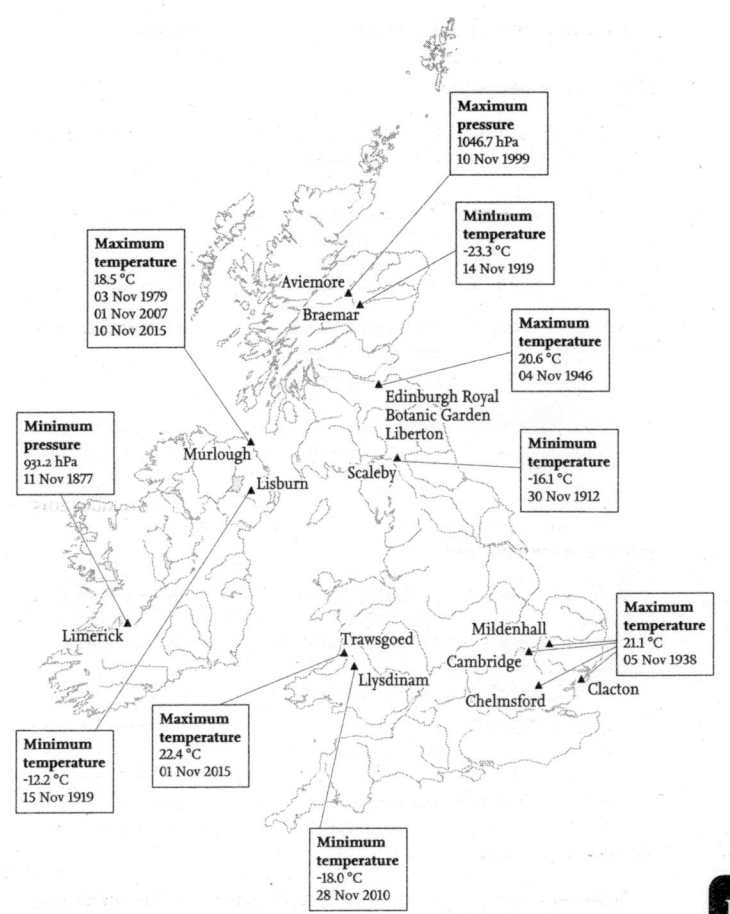

The Weather in November 2024

Observation	Location	Date
Max. temperature 18.8 °C	Treknow (Cornwall)	10 November
Min. temperature −12.4 °C	Kinbrace (Sutherland)	23 November
Most rainfall 126.0 mm	Treherbert (Mid Glamorgan)	23 November
Most sunshine 7.9 hrs	Liscombe (Somerset)	15 November
Highest gust 82 mph (132 kph/71 kt),	Capel Curig (Gwynedd)	23 November
Greatest snow depth 28 cm	Loch Glascarnoch (Ross & Cromarty)	22 November

From mild and dull to cold and wet, November was a month of contrasts. It began where October left off, with high pressure bringing widespread settled conditions. However, a blanket of cloud trapped under the anticyclone led to incredibly dull weather. Thick, low cloud cover persisted for days, known as anticyclonic gloom. By the end of the first week only 2 per cent of average rain had fallen and swathes of southern England, Wales and Ireland hadn't recorded a single hour of sunshine. This wasn't always the case across Scotland, where on the peripheries of the high-pressure system, clouds were able to break, allowing some sunshine through. With southwesterly winds creating a Foehn effect (see page 25), northwestern Scotland saw clear skies and highs of up to 18 °C on 7 November, ten degrees above average for the time of year. It wasn't until late on Remembrance Sunday that conditions began to change. An occluded front introduced a much fresher air mass, clearing the cloud and bringing the first sunshine of the month to many southern areas on Armistice Day. Once

NOVEMBER

skies cleared, a new area of high pressure brought further fine weather. However, from the 15th a cold front moving southwards introduced a blast of cold Arctic air in from the north. Temperatures plummeted, bringing sharp frosts and the first widespread snow of the season. Snow showers piled into northern Scotland and flurries of snow even fell across London. The morning of the 19th saw Braemar, Aberdeenshire, record its lowest temperature this early in the season since 1998, a staggering −11.2 °C. The cold spell came to an end with the arrival of Storm Bert on the 23rd. It brought strong winds and exceptionally wet weather to southwestern England, South Wales and parts of the Midlands, with the UK recording its wettest calendar day since 3 October 2020. Storm Conall arrived a few days later on the 27th. Named by the Dutch meteorological service, it tracked across southern England, bringing heavy rainfall. Strengthening as it moved eastwards, it brought very strong winds to the Netherlands.

With a mild start followed by some colder spells, November's temperatures were close to average. The settled beginning also led to a drier-than-average month, with only 68 per cent of normal rainfall for November. The persistent anticyclonic gloom meant sunshine was also below average, at 89 per cent, but with clearer skies in Scotland it was closer to average.

The averages are:

Maximum temperature	9.42 °C
Minimum temperature	3.56 °C
Rainfall	123.34 mm
Sunshine	57.95 hrs
Air frost days	5.51

Historic Events

2 November 1965 – The Comet computer at the Met Office produced the UK's first operational numerical weather prediction forecast.

10 November 2015 – Storm Abigail became the first storm in the UK to be named by the Met Office. It hit northwestern Scotland several days later, bringing a gust of 84 mph to the Outer Hebrides.

16 November 1771 – A thaw of snow and ice in Upper Teesdale, combined with near-continuous rain, led to widespread flooding in northeastern England. The rivers Tyne, Tees, Wear and Eden burst their banks, destroying bridges and inundating homes with floodwater.

25 November 2005 – Freezing northerly winds combined with a convergence line over Cornwall led to heavy snow across the southwestern county. More than a thousand people were stranded in their cars and thousands of children were stuck at school as snow built through the afternoon. Between 5 and 10 cm of snow fell widely, with up to 30 cm reported on high ground over the moors.

26 November 1703 – A deep and destructive storm, known as the 'Great Storm of 1703', hit central and southern England. Tiles and slates were ripped from roofs and around two thousand chimneys were toppled across the capital city. Further afield, thousands of barns and houses were destroyed, while a storm surge along the Severn Estuary flooded coastal villages of Somerset and the centre of Bristol. Fatalities in England and Wales numbered 123, and even more died in shipwrecks at sea.

NOVEMBER

In this month...

9 November – New Moon

9–20 November – COP31

24 November – Beaver Moon

Look out for: Radiation fog
Similarly to advection fog (see page 89), radiation fog is also named for the process in which it forms. This type of fog is common in autumn and winter and typically forms overnight. Under clear skies and light winds, the ground cools through radiation. This in turn cools the air directly above it, triggering condensation and the formation of fog. When the Sun rises and begins to warm the ground again, the fog usually dissipates.

Morning mist collecting in a valley in Cumbria.

Observing the Weather – Cloud in a Bottle

You will need:
- Clear plastic water bottle
- Water
- A match

To understand how clouds form, the best way is to make one for yourself. First, fill your plastic bottle about two-thirds full with water. Next, light a match, let it burn for a second or two, then drop it into the water bottle and quickly screw the cap on to seal the bottle. Holding it around the middle, squeeze the bottle to increase the pressure inside, then release. As you reduce the pressure within the bottle, a cloud should form!

Why? As well as liquid water in the bottle, there are also particles of water in its gas form – water vapour – which are always present in the air around us. As you squeeze the bottle and increase the pressure, you increase the temperature. As you release it and decrease the pressure, the temperature decreases and some of the water vapour in the bottle condenses. The match is key as it provides cloud condensation nuclei for the water molecules to condense onto. Without that, the experiment wouldn't work.

A plastic bottle that you can easily squeeze to increase the pressure will work best.

Sun and Moon Times in November 2026

Location	Date	Sunrise	Sunset	Moonrise	Moonset
Belfast					
	01 Nov (Sun)	07:26	16:47	22:38	14:21
	11 Nov (Wed)	07:46	16:28	10:51	16:50
	21 Nov (Sat)	08:05	16:13	14:20	03:48
	30 Nov (Mon)	08:20	16:03	23:22	12:55
Cardiff					
	01 Nov (Sun)	07:06	16:45	22:41	13:55
	11 Nov (Wed)	07:23	16:29	10:14	17:04
	21 Nov (Sat)	07:40	16:16	14:16	03:31
	30 Nov (Mon)	07:54	16:07	23:16	12:36
Edinburgh					
	01 Nov (Sun)	07:19	16:32	22:20	14:17
	11 Nov (Wed)	07:40	16:12	10:53	16:25
	21 Nov (Sat)	08:01	15:55	14:06	03:39
	30 Nov (Mon)	08:17	15:44	23:08	12:47
London					
	01 Nov (Sun)	06:53	16:33	22:28	13:42
	11 Nov (Wed)	07:11	16:16	10:02	16:51
	21 Nov (Sat)	07:28	16:03	14:04	03:18
	30 Nov (Mon)	07:42	15:55	23:03	12:24

*all times in GMT

Twilight Times in November 2026

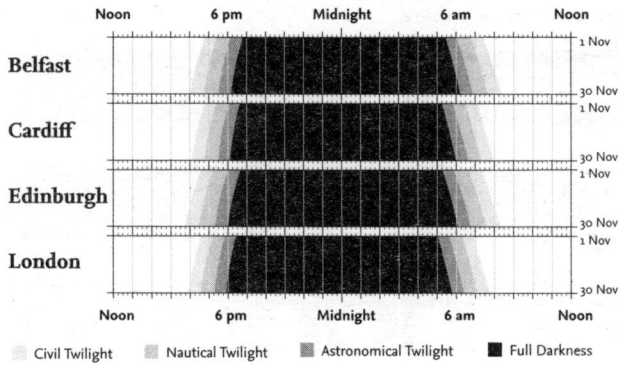

Through November, the decline into darkness remains as steady as ever. Between one and a half and two hours of daylight are lost through this month, depending on location, with long, dark evenings perfect for the lighting of fireworks during the festive celebrations of Diwali and Bonfire Night.

NOVEMBER

Moon Phases in November 2026

Northern Hemisphere

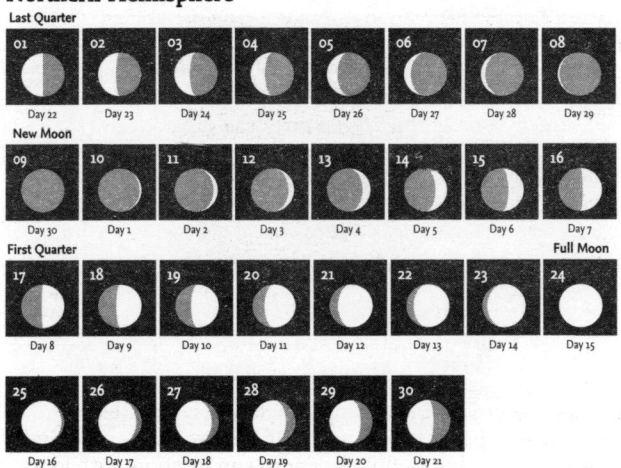

The Beaver Moon
As winter approaches, many animals begin to seek out or create a safe place to hibernate. Beavers start building their dams and it's this process which gives November's Full Moon its name. Another popular name for this moon is the 'Frost Moon', no doubt as this is the time of year that the first widespread frosts appear.

The Jersey Tornado

The Channel Islands are not unfamiliar with extreme weather. Located between 10 and 30 miles off the French coast of Normandy, the islands experience highly changeable conditions: thick fog persisting for days, destructive continental thunderstorms, fierce gales and even snow, despite their generally mild winters. Strong winds bring particular disruption to the islands as flights are grounded and ferries stop running. On the first day of November 2023, the Channel Islands were gearing up to face one such extreme weather event. Named by the Met Office a few days previously, Storm Ciarán was forecast to bring very strong winds to the Channel Islands and the south coast of England. Ahead of its arrival, the Jersey Met weather service issued a red weather warning, and a major incident was declared, closing schools and shops. However, this exceptionally intense weather system brought more than anyone bargained for. As it approached from the west, the storm underwent 'explosive cyclogenesis', deepening by more than 24 hPa within 24 hours. In the late evening of Wednesday, 1 November, showers began to develop along the surface cold front of Storm Ciarán. As it neared the Channel Islands the cluster of showers evolved into a powerful supercell thunderstorm. Then, just before midnight, the UK's strongest tornado in almost 70 years hit the island of Jersey. It created a path of destruction up to 550 metres wide across the east of the island, ripping up around 25,000 trees, tearing roofs off houses and throwing cars and caravans more than 20 metres away from where they had stood. It was rated T6 on the International Tornado Intensity Scale (see page 254), while the accompanying hailstorm, with stones more than 5 cm in diameter, was rated H5 on the TORRO Hailstorm Intensity Scale (see page 257). Away from the tornado, Storm Ciarán brought exceptionally strong winds, damaging enough in themselves. Jersey Airport, on the west side of the island, recorded a gust of 93 mph. It was the highest gust recorded here since the 'Great Storm' of October 1987. Had Storm Ciarán tracked further north, destruction in southern England would have been much more widespread, similar to that of the Great Storm, but instead the focus of the strongest winds was on the Channel Islands and northern France. Storm Ciarán also broke mean sea level pressure records, with the deep low bringing a new lowest pressure to England and Wales for November. Dropping to 953.3 hPa (mb) in Plymouth and 958.5 hPa (mb) in St Athan, both beat their previous records by more than 4 hPa. Following the severe weather, 600 islanders lost power and around 150 people had to seek emergency housing.

Houses destroyed by the Jersey tornado on 1–2 November 2023.

Introduction

As meteorological winter begins, sparkling Christmas lights bring cheer to the wettest and dullest month of the year, December. A lack of sunshine is somewhat expected, as this month holds the winter solstice, the shortest day of the year, when the North Pole reaches its maximum tilt away from the Sun. This moment marks the beginning of astronomical winter. Frequent storms are also to blame for the lack of sunshine and unsettled weather, as they continue to bring strong winds and heavy rain to the UK. This was certainly the case in December 1912, not only the dullest December, but the dullest calendar month on record. The UK averaged a measly 19.4 hours of sunshine through the month, following on from the dullest November on record, at a slightly better average of 33.5 hours. The weather was persistently cyclonic and depressions passed across the UK one after another. Gale-force winds were recorded by at least one weather station on 24 days out of the whole month and Seathwaite in Cumbria recorded 584 mm of rain. The month was characterized by unusually high pressure over the Azores and anomalously low pressure over Iceland. The difference in pressure between these two areas is known as the North Atlantic Oscillation (NAO) and it provides a measure of the strength of westerly winds across the Atlantic. In December 1912, it was particularly strong, known as a positive NAO index, which typically leads to unsettled weather in winter, as it did in this case. When the NAO is weak, or negative, the winter weather tends to be colder and drier, as winds from the north or east become more predominant. The NAO is one of several factors that influence our winters in the UK. The El Niño Southern Oscillation (ENSO) is another such factor that, despite its distance from the UK, can influence our weather through 'teleconnections'. ENSO is a measure of sea surface temperatures in the tropical Pacific Ocean, with El Niño as the warm phase and La Niña as the cold phase. The link between ENSO and our weather is incredibly complicated. However, some scientists suggest that El Niño leads to a wet and windy start to winter in the UK, followed by a cold end, while La Niña is the opposite. Other factors that affect our winters are the stratospheric polar vortex (see page 30) and the extent of

DECEMBER

Arctic sea ice, which can affect atmospheric circulation. The interactions between these factors, not only with our weather but with each other, is very complex and hard to predict. Each must be considered by meteorologists hoping to create a long-range winter forecast, causing quite the headache! Easier to predict, though not by much, is the chance of a white Christmas, which begins to attract interest in the first days of December. Forecasts for the month ahead can provide an idea of whether it will be warmer or colder, wetter or drier, but not until Christmas is much closer can forecasters say with any certainty if it will be white. Technically speaking, only one snowflake needs to fall somewhere in the UK on the 24 hours of Christmas Day for it to be a white Christmas, so the chances are higher than you'd think!

Weather word: Flaggie (Scottish and Northern English, nineteenth century)
A large snowflake, as in: 'a fall of soft, flaggie snow'.

WEATHER ALMANAC 2026

Weather Extremes in December

Country	Temp.	Location	Date
Maximum temperature			
England	17.7 °C	Chivenor (Devon)	2 December 1985
		Penkridge (Staffordshire)	11 December 1994
Wales	18.0 °C	Aber (Gwynedd)	18 December 1972
Scotland	18.7 °C	Achafry (Sutherland)	28 December 2019
Northern Ireland	16.7 °C	Ballykelly (County Londonderry)	2 December 1948
Minimum temperature			
England	−25.2 °C	Shawbury (Shropshire)	13 December 1981
Wales	−22.7 °C	Corwen (Denbighshire)	13 December 1981
Scotland	−27.2 °C	Altnaharra (Sutherland)	30 December 1995
Northern Ireland	−18.7 °C	Castlederg (County Tyrone)	24 December 2010

Country	Pressure	Location	Date
Maximum pressure			
Scotland	1051.9 hPa	Wick (Caithness)	24 December 1926
Minimum pressure			
Northern Ireland	927.2 hPa	Belfast (County Antrim)	8 December 1886

DECEMBER

Minimum temperature
−27.2 °C
30 Dec 1995
— Altnaharra

Maximum pressure
1051.9 hPa
24 Dec 1926
— Wick

Maximum temperature
18.7 °C
28 Dec 2019
— Achfary

Maximum temperature
16.7 °C
02 Dec 1948
— Ballykelly

Minimum temperature
−18.7 °C
24 Dec 2010
— Castlederg

Minimum pressure
927.2 hPa
08 Dec 1886
— Belfast

Maximum temperature
18.0 °C
18 Dec 1972
— Aber

Minimum temperature
−22.7 °C
13 Dec 1981
— Shawbury

Minimum temperature
−25.2 °C
13 Dec 1981
— Corwen

Maximum temperature
17.7 °C
02 Dec 1985
— Chivenor

Maximum temperature
17.7 °C
11 Dec 1994
— Penkridge

D

The Weather in December 2024

Observation	Location	Date
Max. temperature 15.9 °C	Westonzoyland (Somerset)	1 December
Min. temperature −11.2 °C	Tyndrum No. 3 (Perthshire)	11 December
Most rainfall 144.2 mm	Honister Pass (Cumbria)	31 December
Most sunshine 6.9 hrs	Exeter Airport No. 2 (Devon)	8 December
Highest gust 96 mph (154 kph/83 kt)	Berry Head (Devon)	7 December
Greatest snow depth 5 cm	Strathy East (Sutherland) and Loch of Hundland (Orkney)	31 December

A very mild, cloudy month with frequent unsettled weather, the first week of December saw depressions interspersed with transient ridges of high pressure. Storm Darragh arrived on the 6th and a rare red severe weather warning for strong winds was issued by the Met Office. The second of only two issued in 2024, it covered western Wales and areas surrounding the Severn Estuary. Here, gusts of 70–90 mph were recorded, while the strongest gust, 96 mph, was in Devon. As Storm Darragh cleared, high pressure developing over Scotland brought a quieter week of weather, although brisk northeasterlies led to a marked wind-chill across England and Wales. Skies were cloudy here, while Scotland and Ireland saw the best of the winter sunshine. Under high pressure on the 11th, a temperature inversion brought sub-zero daytime highs in the valleys and glens but warm summits: 6.8 °C at Cairngorm and 7.1 °C at Cairnwell. By mid-month, milder conditions and breezy

DECEMBER

westerlies arrived with Atlantic depressions. One of these, tracking to the north of the UK, brought a wet and windy winter solstice, then introduced another spell of cold weather as the low cleared. High pressure built from the 23rd, bringing a settled and mild festive period. Anticyclonic conditions saw the return of persistent low cloud, mist and fog. Christmas Day was generally dry, mild and cloudy, with Scotland and Ireland recording their mildest starts to the big day on record, at 11.0 °C and 11.9 °C, respectively. It was the mildest Christmas since 2016 and with no snow falling in the UK, it was not a white Christmas. Under a blanket of fog over the following week, some weather stations did not record a single hour of sunshine. In the last few days, high pressure began to retreat southwards. Conditions turned unsettled on the 31st, with a depression bringing widespread wet and windy weather.

Thanks to frequent mild westerlies, December was 2 °C warmer than average across the UK, making it the fifth warmest on record. It was also wetter than usual, with 110 per cent of the average rainfall. Persistent fog at the end of the month had a big impact on sunshine hours in the UK, which ended up at only 57 per cent of average – equal to 24.3 hours. Averaged across the whole month, that equates to less than one hour of sunshine per day. It was also the fourth-dullest December on record.

The averages are:

Maximum temperature	7.02 °C
Minimum temperature	1.42 °C
Rainfall	127.16 mm
Sunshine	42.68 hrs
Air frost days	10.95

Historic Events

8 December 1954 – A violent tornado ripped the roof off Gunnersbury London Underground station, injuring eight people who were waiting on the platform. The T7 tornado tracked 16 km across West London, ripping down walls in Acton, Chiswick and Willesden, too. It was accompanied by active thunderstorms, with frequent lightning, heavy rain and hail.

12 December 2015 – On the last day of COP21, the landmark Paris Agreement was reached, aiming to limit global temperature rise to 1.5 °C above pre-industrial levels.

24 December 1997 – The 'Christmas Eve Gale' brought strong winds to west Wales and northwestern England, with gusts of 70–80 mph widely, reaching 110 mph at Aberdaron on the Lleyn Peninsula. Uprooted trees blocked roads and fallen power lines left 50,000 homes without electricity on Christmas Day.

25 December 2010 – The last widespread white Christmas in the UK, with snow on the ground at 83 per cent of weather stations and falling sleet or snow at 19 per cent of stations.

27 December 1836 – The deadliest avalanche in British history occurred in Lewes, East Sussex. It followed a spell of very cold and snowy weather, with a powerful snowstorm on Christmas Eve creating huge drifts. A deep layer of snow built up on the sheer edge of Cliffe Hill, overhanging the houses below. It inevitably collapsed on the 27th, destroying the cottages at the foot of the hill and killing eight people.

DECEMBER

In this month...

1 December – Meteorological winter begins

9 December – New Moon

21 December – Astronomical winter begins

21 December – Winter Solstice

24 December – Super Cold Moon

Look out for: Diamond dust
Like millions of tiny gems glittering in mid-air, diamond dust is the name given to slowly falling ice crystals in clear, calm and very cold weather. Sparkling in sunlight, they typically form in temperatures below -10 °C as water vapour in the air freezes. Each individual ice crystal is so small it can't be seen with the naked eye, but together they create a beautiful shimmering effect.

Observing the Weather – Nacreous Clouds

While midsummer skies hold the promise of a possible noctilucent cloud display, midwinter can bring the chance of another rare, upper atmosphere cloud sighting. Nacreous clouds are known for their distinctive 'mother-of-pearl' sheen and, like noctilucent clouds, are illuminated by the Sun from beneath the horizon. Also known as polar stratospheric clouds, they are usually confined to the polar regions, but can be glimpsed from the mid-latitudes if the air in the stratosphere is cold enough, as nacreous clouds require temperatures of around −80 °C in order to form. We are more likely to see them in the winter, when the lack of sunlight causes temperatures in the stratospheric polar vortex to plummet. They are composed of a mixture of nitric acid, sulphuric acid and ice crystals, the latter forced upwards into the stratosphere by the orographic waves responsible for lenticular clouds (see page 240). Their iridescent pastel hues are formed by the scattering of light as it passes through the ice particles, similar to the way in which a rainbow is formed.

In the UK, they can usually only be seen when the stratospheric polar vortex is displaced from its home above the North Pole and hovers briefly overhead. The likeliest time to see them is at sunrise or sunset during winter.

Despite their whimsical appearance, nacreous clouds are damaging to the ozone layer – part of the atmosphere that protects us from the Sun's ultraviolet radiation. This is because their chemical composition encourages a reaction within the ozone layer, speeding up its depletion.

DECEMBER

Nacreous clouds at dawn in Golspie, Scottish Highlands, January 2023.

WEATHER ALMANAC 2026

Sun and Moon Times in December 2026

Location	Date	Sunrise	Sunset	Moonrise	Moonset
Belfast					
	01 Dec (Tue)	08:22	16:02	–23:22	13:04
	11 Dec (Fri)	08:35	15:57	11:02	17:42
	21 Dec (Mon)	08:44	15:59	13:13	05:49
	31 Dec (Thu)	08:46	16:07	01:15	11:39
Cardiff					
	01 Dec (Tue)	07:55	16:07	–23:16	12:50
	11 Dec (Fri)	08:08	16:03	10:29	17:52
	21 Dec (Mon)	08:15	16:05	13:21	15:19
	31 Dec (Thu)	08:18	16:13	00:58	11:35
Edinburgh					
	01 Dec (Tue)	08:19	15:44	–23:08	12:55
	11 Dec (Fri)	08:33	15:38	11:02	17:19
	21 Dec (Mon)	08:42	15:39	12:53	05:47
	31 Dec (Thu)	08:43	15:47	06:06	11:24
London					
	01 Dec (Tue)	07:43	15:54	–23:03	12:38
	11 Dec (Fri)	07:55	15:51	10:16	17:39
	21 Dec (Mon)	08:03	15:53	13:08	05:07
	31 Dec (Thu)	08:06	16:00	00:45	11:22

*all times in GMT

Twilight Times in December 2026

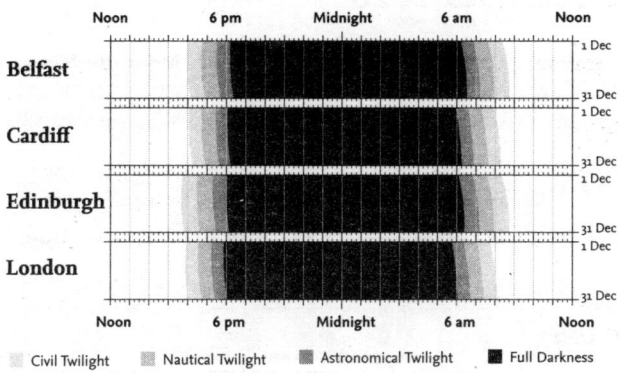

The darkest month of the year is here, culminating with the winter solstice on the 21st. On this day, parts of northern Scotland only receive six and a half hours of daylight, while those in southwestern England get closer to eight hours. However, the earliest sunset happens around a full week before the solstice, depending on where you live – in London for example, this is around the 12th/13th. Thereafter, evenings begin to get ever so slightly lighter. However, to make up for this, our sunrises continue to get later, even after the winter solstice, until just before the New Year.

Moon Phases in December 2026

Northern Hemisphere

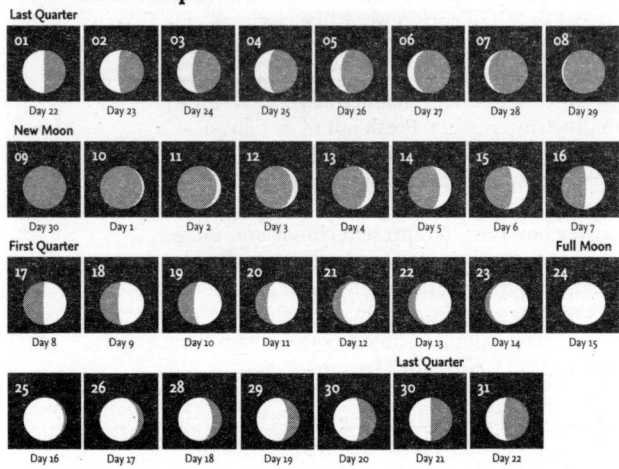

The Cold Moon
Rather than a warning about the weather, December's Super 'Cold Moon' is named to mark the beginning of winter. In 2026, it's also a supermoon, since the Moon is at 'perigee', the closest point to Earth in its orbit. This makes it appear bigger and brighter in the sky.

Other names for December's Full Moon include 'Moon before Yule' or the 'Long Night Moon', both referring to the winter solstice.

Professor Dame Julia Slingo OBE

What do you see when you look out of the window? For many of us, a sky full of cotton wool-like cumulus clouds is just that, but Julia Slingo sees physics in action. Back in the 1970s, it was her love of physics and the weather that led to her joining the Met Office, a moment that defined the trajectory of her long and illustrious career. Fresh out of her Physics degree at Bristol University, Julia's first job as a Met Office scientist was working on pioneering climate models. At the time, she was one of only a handful of women doing research. Her particular role was to discover how best to represent clouds and the way they interact with the atmosphere. Being closely linked to this, Julia also investigated how increasing levels of carbon dioxide would affect our planet. In the early 1980s she left the Met Office to have her first daughter, then rejoined the world of weather at the European Centre for Medium-Range Weather Forecasts (ECMWF). Julia stayed here for around five years before having a second daughter and moving across the pond to join the US National Center for Atmospheric Research, where she researched intraseasonal variability in the tropics and the role of clouds within these systems.

Julia returned to the UK in the 1990s, where she took up the post of Research Fellow at the University of Reading and continued her work on climate models. Ten years later, she became the first female professor of meteorology in the UK. Following this achievement, Julia took on the role of Director of Climate Research at the National Centre for Atmospheric Science, based at Reading University, pushing the boundaries of high-resolution climate modelling. She also founded the Walker Institute for Climate System Research. Time and time again, Julia was recognized for her outstanding contributions in science, and continued to break glass ceilings with her achievements. In 2008, she became the first female president of the Royal Meteorological Society and a year later was headhunted for the role of Chief Scientist at the Met Office, the first woman to do that, too. Overseeing everything from the running of day-to-day weather forecast models to complicated climate simulations, and helping them work seamlessly together, Julia brought one innovation after another to the Met

Office. Knowing how it felt to be the only woman sitting at a table full of men, she helped encourage her female colleagues to recognize their own potential too. As is often the case for those working in weather, she also dealt with the unexpected, such as the eruption of the Icelandic Eyjafjallajökull volcano in 2010. Receiving the recognition she deserves, Julia was awarded an OBE in 2008 and made a Dame in 2014 for her services to weather and climate science. She left the Met Office in 2016 and although now retired, she still maintains the love for physics that drew her to the weather in the first place.

Professor Dame Julia Slingo OBE.

WEATHER ALMANAC 2026

ADDITIONAL INFORMATION

Our Regional Climates

Introduction
As anyone living in these islands knows, our weather is incredibly variable. However, there is a difference between weather and climate, the latter being the weather averaged over a long period of time. According to the Köppen climate classification system, which categorizes the world into 30 different climate types, the British Isles have a 'humid temperate oceanic' climate. This is thanks to the influence of the Atlantic Ocean, whose proximity gives us warmer weather than is typical of other places of a similar latitude, and also brings plenty of moisture. Meanwhile, Europe is still close enough to provide a continental influence from time to time.

However, across the UK, Ireland and crown dependencies, the climate from one region to another can exhibit a marked difference. This is due to a variety of reasons, including elevation, the distribution of mountain ranges, proximity to the coastline, and latitude, to name a few. Overall, the climate can be described as being warmer and drier in the south and east, while conditions are generally cooler and wetter the further north and west you go. With this in mind, the British Isles can be split into eight distinct areas, each with its own regional climate. They are as follows:

    Southwest England and the Channel Islands
    Southeast England and East Anglia
    The Midlands
    Northwest England and the Isle of Man
    Northeast England and Yorkshire
    Wales
    Scotland
    Ireland

Southwest England and the Channel Islands
Fascinating geographical landscapes in the southwest make for equally fascinating weather. With expansive, wind-scoured moorlands, rolling hills and deep, winding valleys, the climate can be very variable from one place to another, with the unique topography lending itself to the formation of

WEATHER ALMANAC 2026

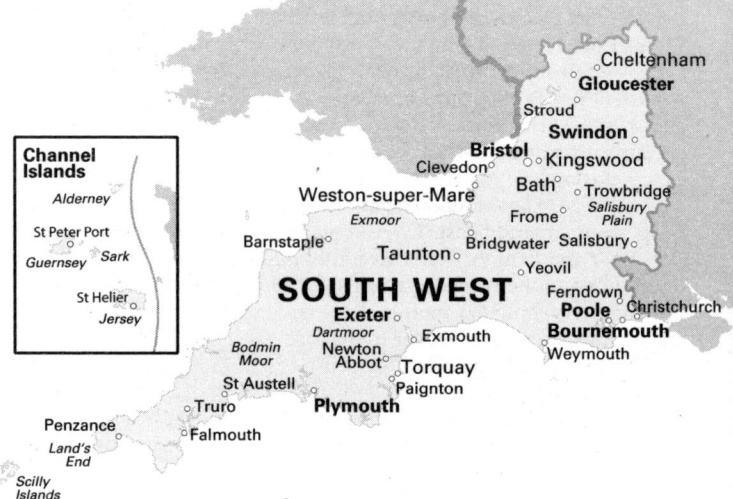

local microclimates. Not to mention the ever-present coastal influence, to which the southwest owes its mild, maritime climate. The stunning scenery and pleasant climate support a healthy tourism industry, with a strong agricultural sector too. This region stretches from Cornwall, in the far southwest, to Gloucestershire in the north and Dorset in the east, while also including the Channel Islands and Isles of Scilly. Both of the latter pride themselves on their warm and sunny climates. There are no hills on Scilly to force orographic rainfall, so these islands also have a relatively dry climate. Meanwhile, the Channel Islands are heavily influenced by wind direction. Winds from the east bring the continental influence of nearby France, while mild, moist air from the west can often lead to persistent fog. Both groups of islands are too small to generate land-based convection during the summer months, although occasionally some fair-weather cloud will develop over the Channel Islands.

While the island archipelagos are generally rather low-lying, the mainland boasts three large moors (Bodmin Moor, Dartmoor and Exmoor) and three hill ranges (the Mendips, the Cotswolds and the Blackdown Hills). The highest of these is Dartmoor, reaching 621 metres above sea level, high enough to cast a rain shadow over Exeter to the east.

The seas which surround the southwest peninsula have the highest annual mean temperatures of any around the UK. Even in winter and early spring, when they're at their coolest, they still keep coastal areas relatively mild, making the southwest the warmest place in the country at this time of year. The Severn Estuary has a similar wintertime effect, transporting milder air towards the heart of England as far as the southwest Midlands.

Away from the mild seas and estuaries, altitude has more of an influence on temperature, with towns and villages on the moors colder than their lower-altitude neighbours.

These higher-ground areas also receive the greatest rainfall. While most western areas (i.e. Cornwall and Devon) see annual rainfall somewhere close to 1,000 mm, the moors typically measure double this. Meanwhile further east, rainfall totals are lower, with the Somerset Levels at the other end of the extreme, receiving less than 800 mm a year on average. For all of these areas, rainfall is highest during autumn and winter, and lowest during the spring and summer. It's during these warmer months that the proximity of the southwest to the Azores high helps to bring the best of the settled weather and sunshine, especially along the south coast. However, the region is also known for heavy rainfalls at this time of year. Martinstown in Dorset holds the record for the highest 24-hour rainfall total, recorded from 9 a.m. on 18 July in 1955. Severe thunderstorms brought a deluge of 279 mm to the small village, with around 190 mm believed to have fallen within five hours. The Lynmouth and Boscastle (see page 131) events are also examples of extreme convective rainfall, which led to devastating flooding. The short, steep valleys that are a hallmark of the Cornwall and Devon countryside are notorious for flash flooding and as our climate changes, this is likely to worsen.

Southeast England and East Anglia
A region of vast skies, leading to spectacular cloudscapes, and rich soils which support its substantial agricultural industry, the southeast boasts the warmest, driest, sunniest and calmest climate within the UK. Furthest away from the average path of Atlantic depressions, this area receives the greatest influence from the nearby continent, which brings hot, humid weather in the summer but can also lead to chilly winters. The weather experienced in the southeast is broadly similar across the whole region, but the area can be further divided into two distinct sub-regions: those to the north of London (East Anglia and the Home Counties) are driest while those to the south (Hampshire, West and East Sussex, Kent and the Isle of Wight) are slightly wetter. The Isle of Wight enjoys a similar climate to the nearby coast, but its exposed nature often leads to strong winds during winter storms. In fact, the weather station at Needles Old Battery on the far western tip of the island holds the record for highest low-level wind gust in England, reaching 122 mph during Storm Eunice in 2022.

In the summer, the southeast is typically the warmest place to be in the UK. Its proximity to the Continent means it spends more time in warm tropical air masses than regions

ADDITIONAL INFORMATION

further north, for which the incursions of tropical air tend to be much more short-lived. However, the hundreds of miles of coastline help to take the edge off the heat in many places, with sea breezes a common feature of late spring and summer. On the other hand, the relatively mild seas surrounding the region in the winter often help coastal areas stay much milder. Meanwhile, temperatures plummet across inland areas, with the southeast being home to several well-known frost hollows, including Santon Downham in Suffolk and Benson in Oxfordshire. At both of these, only July and August stay free of air frosts on average, while grass frosts can occur in any month.

In winter, chilly northerlies can bring cold air and frequent snow showers to Norfolk, but the most severe winter weather here tends to come from the east. Freezing polar continental air, often originating from the heart of Siberia, approaches the UK over the North Sea. Here, it picks up plenty of moisture and upon meeting the southeast, drops it all in the form of heavy snow, especially across East Anglia. The 'Beast from the East' in 2018 is the most memorable example of this from recent years.

Home to some of the driest places in the UK, the southeast tends to get the tail-end of the rain as weather fronts move across the country. This is thanks to a 'rain shadow' effect, whereby most of the rain from a system falls over higher ground in the west of the UK. Therefore, winter precipitation totals here are typically lower than the rest of the country. However, it makes up for it in summer, as the southeast becomes the home of thunderstorms. Developing here or imported from the nearby continent, these dramatic displays bring frequent lightning, torrential rainfall and even a chance of tornadoes. It's this boost in convective rainfall during the summer months that gives the southeast one of the most even distributions of rainfall through the seasons in the UK.

Despite its somewhat benign reputation, the southeast is the most vulnerable to climate change. It's warming at a greater rate than anywhere else in the UK, with its already hot summers getting even hotter and the chance of drought increasing too, putting further pressure on the region's agriculture. Meanwhile, with so much of East Anglia's land lying below sea level, it's also at risk from coastal flooding and rising seas.

The Midlands
Lying in the geographic heart of England, the Midlands' most defining climate characteristic is its range in temperatures. Away from the heating influence of the sea in winter and its cooling effect in summer, the Midlands often records the highest maxima and lowest minima in England. In fact, it holds the record for the highest temperature in the UK, 40.3 °C at Coningsby on 19 July 2022 (see page 116), as well as the lowest on record in England, −26.1 °C on 10 January 1982 in Shropshire (see page 20). The region is predominantly low-lying, but with notable hills around its boundaries, including the foothills of the Welsh mountains, the Staffordshire moorlands, the Peak District and the Northamptonshire Jurassic escarpment. In the centre of these is the Birmingham Plateau, surrounded by three valleys: the Avon, the Severn and the Trent. The eastward extent of this region also includes Lincolnshire.

This somewhat landlocked region has an almost continental climate, as far as that can be true on an island such as Great Britain. Without the sea to moderate temperatures, the Midlands responds to changes in radiation faster than other regions. In winter, temperatures are highest in the southwest, thanks to warmth transported inland from the Severn Estuary; this is the extent of the effect the sea has on the climate in the Midlands. Otherwise, it can become very cold, especially at night. With its sandy soils which cool quickly, the Midlands is particularly prone to frosts. In the west, cold air drains into the river valleys of the Severn, Avon and Wye, leading to very chilly nights for the surrounding areas. Meanwhile in summer, the highest temperatures are generally found in the eastern Midlands. While western areas tend to be cloudier, thanks to high ground and mild, moist air arriving from the Atlantic, eastern areas tend to experience clearer skies, with fewer clouds to impede the radiation from the Sun.

WEATHER ALMANAC 2026

This is effectively thanks to the rain shadow cast across the Midlands by the high ground in Wales, which affects rainfall distribution in the same way. The west of the region is much drier than it would be if Wales had no mountains, but is still the wetter half of the Midlands, especially near the Welsh border and over the Peak District. This area sees more rainfall in autumn and winter than spring and summer. Travelling eastwards, conditions tend to become drier, with the exception of higher-ground areas like the Birmingham Plateau. Here, there is an annual difference of 100 mm of rain between Birmingham and Coventry, which is situated at a lower level. While eastern areas are generally drier, they tend to experience more thunderstorms and convective rainfall than in the west. Occurring in spring and summer, these tend to even out the annual rainfall distribution, compared to the west with its distinct peak in autumn/winter. While the Midlands is generally sheltered from showers moving inland on northerly winds, there are occasions when they can sneak in. During a northwesterly airflow, showers developing over the Irish Sea are funnelled through the Cheshire Gap towards Birmingham, even reaching as far as London at times. Meanwhile, northeasterly winds can drive North Sea showers into Lincolnshire, and when cold enough in winter, these can bring significant snowfall accumulations.

Northwest England and the Isle of Man

With the Irish Sea stretching out to the west and the Pennines soaring to the east, northwestern England is home to a variety of contrastingly beautiful landscapes. From the craggy fells and azure waters of the Lake District, home of England's highest peak, to the undulating countryside of the Cheshire Plains, the diverse geography of this region leads to a fairly varied climate.

Despite local topography leading to a few relatively dry areas, the region has a reputation for being wet, with mild winters and cool summers. This made it an excellent setting for the historic cotton industry, which relied on high humidity for successful spinning. Lying offshore to the west of Cumbria, roughly equidistant from the coasts of southern Scotland,

WEATHER ALMANAC 2026

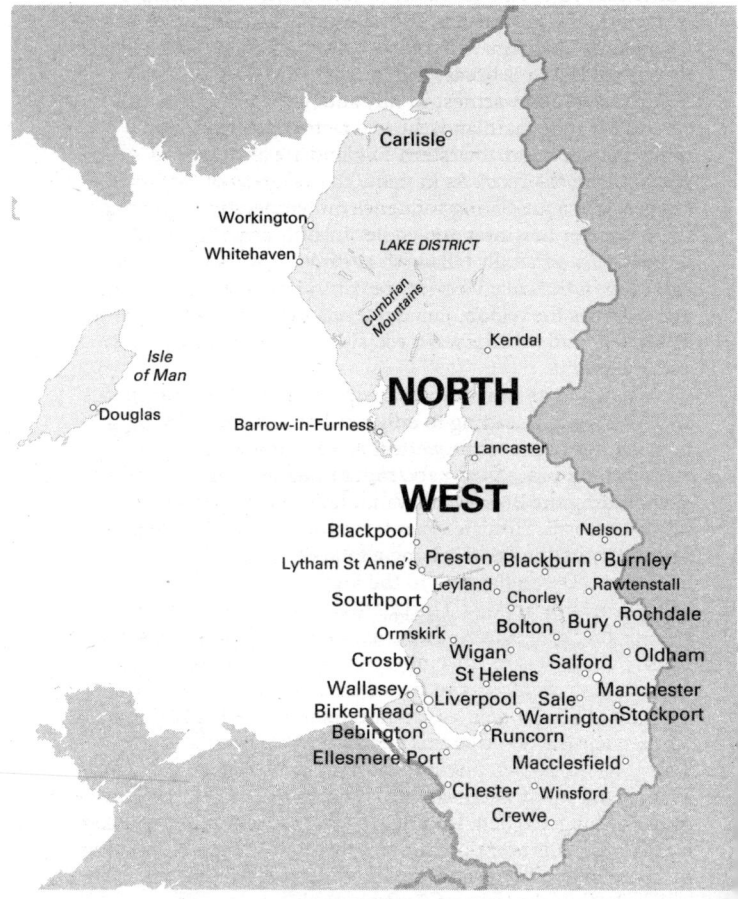

ADDITIONAL INFORMATION

Northern Ireland and northwestern England, is the Isle of Man. Rocky coastlines rise gradually to rolling hills, culminating in a central range of peaks, the tallest of which is Snaefell, at 621 metres. The climate here has a heavy maritime influence, thus is coldest in February, when seas are at their lowest temperature, and warmest in July and August.

Back on the mainland, the two factors determining temperatures in northwestern England are altitude and proximity to the coast. As in many coastal areas of Britain, the sea provides a mediating influence on temperature, keeping them warmer in winter and cooler in summer. Meanwhile, temperature generally falls with altitude so areas with high elevation in the Lake District are naturally cooler than lowland areas. Across the region, January tends to be the coldest month of the year and July the warmest, similar to much of the UK and Ireland.

Thanks to its proximity to the path of Atlantic depressions and high ground leading to enhanced orographic rainfall, the northwest is home to the wettest place in England. Receiving an average of more than 3,200 mm of rain annually, the fells surrounding the Borrowdale valley hold many of the UK's rainfall records. Honister Pass takes the top spot for highest 24-hour rainfall total, receiving a total of 341.4 mm of rain from 6 p.m. on 4 December 2015 to the same time the following day. The nearby areas of Thirlmere and Seathwaite also hold records for highest two-, three- and four-day totals measured from 9 a.m. It was in November 2009 that the latter of these extreme rain events was recorded, as a moist southwesterly airstream brought near-constant rainfall to the Lake District. Some scientists describe this effect as an 'atmospheric river'. This is a well-known phenomenon in North America, whereby a large amount of water vapour is transported along a relatively narrow, elongated belt, leading to heavy rainfall when it makes landfall. On this occasion, it not only brought record-breaking measurements, but also severe flooding and devastation, particularly in the town of Cockermouth.

Elsewhere in the region, where elevation is much lower, exist pockets of a significantly drier climate. The Cheshire

Plains and the Vale of Eden lie in the rain shadows of the Welsh mountains and the Lake District, respectively, with some areas here receiving less than 800 mm of rain per year.

Across the northwest, the driest season is spring. Prevailing winds often shift to the northeast and the Helm Wind can develop (see page 66). Meanwhile, the wettest time of year is autumn and winter, thanks to the storms that are frequent during these seasons. As the climate continues to warm and extreme rainfall increases in turn, there are concerns for how the northwest, the wettest place in England, will fare.

Northeast England and Yorkshire
From wide moors blanketed in heather to windswept coastlines, this region is surrounded by four distinct boundaries: the rivers Tweed and Humber to the north and south, the Pennines to the west and the North Sea to the east. It is the two latter in particular that dictate the climate in the northeast. The Pennines are responsible for creating a rain shadow across the region, while the coldest waters in the UK are found off the coast of northeast England, bringing a cool climate all year round. The topography is generally highest in the west, gradually lowering towards sea level in the east, but with a few exceptions, such as the North York Moors and Yorkshire Wolds. Meanwhile, the Cheviots lie in the far northwest.

As is the case for much of the UK, maximum temperatures in the northeast are found in July and minimum temperatures tend to occur in January. Although coastal regions often experience their minima later in the winter, when seas are at their coolest, this is not the case here as prevailing southwesterly winds tend to bring a milder influence, blowing air cooled over the sea offshore. On occasions where winds do turn northerly or easterly in winter, this is when temperatures plunge across the region and snowfall is most likely. Meanwhile in summer, sea temperatures only rise to around 15 °C on average and there can often be a marked increase of temperature from coastal areas to further inland, sometimes rising by 10 °C over just a few kilometres.

ADDITIONAL INFORMATION

The cool seas and northeasterly winds in late spring and summer also bring another phenomenon for which this region is known: fret. Relatively warm, moist air blows over cool seas, causing water vapour to condense into fog. This can linger all day in some instances, bringing dull and chilly weather to coastal areas while conditions inland are warm and sunny.

The combination of prevailing west-to-southwesterly winds with the Pennines and the resultant rain shadow means the northeast is one of the drier regions of the UK. Not only does the high ground force rising air to release its moisture, it also breaks up cloud cover, meaning this region can be rather sunny too. The wettest time of year corresponds with the autumn/winter peak in storms, while spring tends to be drier. Although the northeast doesn't experience convective rainfall to the same scale as the south, showers and thunderstorms are no less severe when they do occur, exacerbated by sea breezes and urban heat island effects. Meanwhile, unstable air masses which arrive from the north in winter can bring frequent hail at this time of year.

Wales
Lush, green and mountainous, Wales has an undeniable reputation for being wet. This can, in part, be attributed to its topography, with a long spine of mountains running from north to south, the highest of which is Yr Wyddfa (Snowdon), sitting at 1,085 metres in the north. Wales's maritime influence and its proximity to low-pressure systems arriving from the Atlantic also contribute to its wetness. Most places in Wales receive more than 1,000 mm of rainfall on average per year, while Eryri (Snowdonia) rivals the wettest parts of the Lake District or the Scottish Highlands, measuring an annual average of more than 3,000 mm of rain. It's no surprise, therefore, that Wales holds the record for the wettest calendar month, December 2015, when Crib Goch, a station just to the east of Yr Wyddfa's peak, recorded 1,396.4 mm of rain. But that's not to say that the whole of Wales is wet. Indeed, thanks to its varied landscape from coast to lowlands, it displays a range

ADDITIONAL INFORMATION

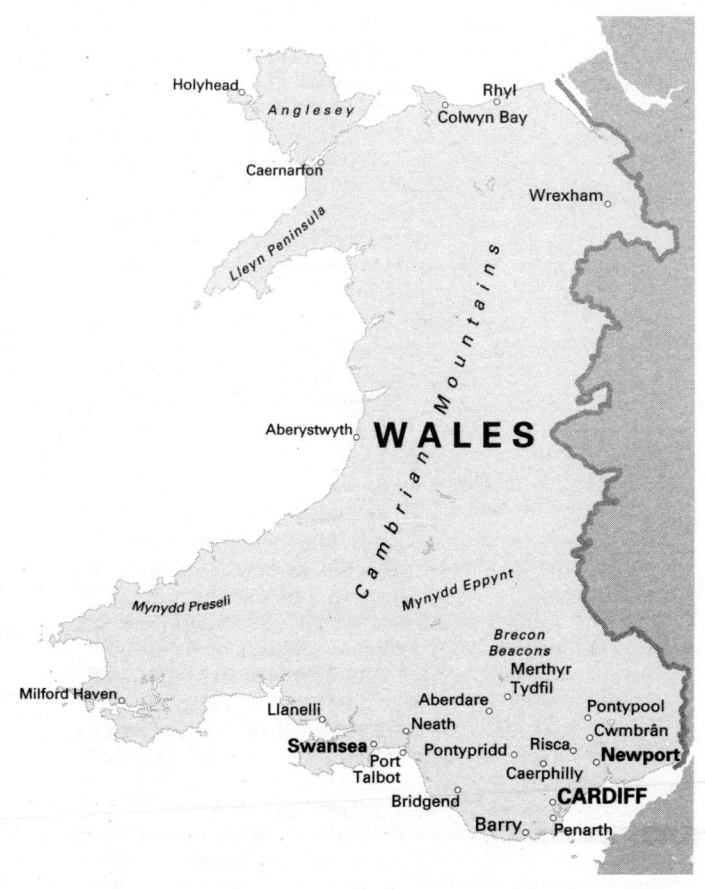

of climates and can therefore be considered as a multitude of sub-regions.

The driest of these is the northeastern lowlands, hidden in the rain shadow of the mountains to the west and with a similar climate to the nearby Cheshire Plain. It's often one of the warmest places in Wales too, with the Hawarden weather station well known for recording high temperatures. In the extreme heatwave of July 2022, Hawarden recorded a maximum of 37.1 °C, the highest temperature ever recorded in Wales. The Foehn effect (see page 25) often plays a part in the record highs here, as Hawarden is in the lee of high ground to the southwest.

Meanwhile, when the wind is blowing across Wales from the east, the Foehn effect can bring high temperatures to another sub-region: the coastal area from St David's Head, round Cardigan Bay to Anglesey. In this set-up, which is common in spring, Aberystwyth and Porthmadog on the west coast can see temperatures soaring in the morning before tailing off into the afternoon as sea breezes kick in. Otherwise, this area displays a very maritime climate: mild in the winter, cooler in the summer and often windy.

Further inland over the mountain ranges of Wales, the climate becomes dull, wet and windy. Orographic uplift causes weather systems to release their rainfall over these areas and also brings cloudy skies, especially during autumn and winter, when the frequency of storms is highest. However, inland areas also see the lowest temperatures, especially when skies are clear and snow is lying on the ground. This was the case when Wales recorded its lowest minimum on record, -23.3 °C on 21 January 1940. It happened at Rhayader in Powys as cold air drained into the valley floor where this decommissioned weather station was located.

To the east of the mountainous spine of Wales, the rain shadow begins to take effect and during westerlies, these areas tend to be drier. However, it's still relatively wet, especially during cyclonic situations. Meanwhile, in easterlies, these areas can be persistently cloudy as stratus 'piles up' along the foothills of these mountains, while west Wales gets the best of the sunshine.

ADDITIONAL INFORMATION

The final sub-region is south Wales, from St David's Head eastwards towards the Severn Estuary. Here it's mild, occasionally hot in summer, especially near Cardiff, which benefits from its proximity to southern England. Easterlies also bring warmth from the southeast to these areas in summer. The coasts here are particularly sunny, as they're free from the fair-weather cloud building inland during spring and summer, as well as the orographic cloud developing over high ground.

As a whole, Wales typically experiences fewer severe thunderstorms and snowfall events than elsewhere in the UK. However, convergence lines forming in northerly airflows can bring frequent showers, falling as snow if the air is cold enough. A well-known example is the 'Pembrokeshire Dangler', a convergence line which runs through the Irish Sea and across the peninsula for which it is named. Meanwhile, extreme snow events are not unheard of, for example in the winter of 1982 (see page 20).

Scotland

Scotland is home to some of the most breathtaking landscapes in the world. In the west, dramatic mountains cleaved by deep lochs give way to hundreds of unique islands reaching out into the Atlantic. Thousands of miles of coastline bring everything from magnificent cliffs to spectacular shorelines. In summer, sweeping white sand beaches and turquoise seas have an almost tropical feel to them – until you take a dip in the sea! In the heart of Scotland, the tallest mountains in the UK tower over shaded glens, while further south they ease into the rolling hills of the lowlands. Naturally, this varied geography has a dramatic impact on Scotland's climate. The islands and exposed western coastlines have exceptionally mild weather, considering how far north they are. Rising steeply from the west coast and extending inland to the heart of Scotland, the Highlands bring strong orographic enhancement, with frequent rain as well as plenty of snow in winter. In the rain shadow to the east, the Sun shines on quaint fishing villages dotted along a coastline of rocky cliffs punctuated by windswept beaches.

In part, Scotland owes its reputation for extreme temperatures to its mountainous landscape, experiencing

WEATHER ALMANAC 2026

incredible cold as frigid air is funnelled into the deep valleys, and surprising warmth, thanks to the Foehn effect (see page 25). Being further north than anywhere else in the UK, Scotland also sees longer days in summer and longer nights in winter, with the limited sunlight in winter helping to exacerbate the extreme cold. On three separate occasions, temperatures have fallen to −27.2 °C, the lowest ever recorded in the UK. Braemar, a village in the heart of the Cairngorms, nestled in the valley of the River Dee, holds two of these records – the first in February 1895 and the second in January 1982 (see pages 20 and 36). More recently, Altnaharra in Sutherland, northern Scotland, recorded the record-equalling low in December 1995. Generally, December and January tend to bring the coldest weather to Scotland. However, those places falling under a maritime influence, such as the Hebrides, Orkney and Shetland, usually see their winter minima in February, when the surrounding seas are at their coldest. Similarly, in summer, the islands are at their warmest at the same time as the seas, in August. Meanwhile, the rest of Scotland generally records summer maxima in July. The Scottish Borders hold the record for the highest temperature measured in Scotland: 34.8 °C on 19 July, during the unprecedented heatwave of summer 2022. Although a far cry from the aforementioned record, unexpected winter warmth is also possible, thanks to the Foehn effect. In fact, Scotland holds the record for highest December and January temperatures, both recorded in Achfary, Sutherland. This mountain weather phenomenon led to a high of 18.7 °C in December 2019 and, more recently, 19.9 °C in January 2024.

The mountains are also responsible for the distribution of rainfall across Scotland, which shows a strong west-to-east gradient. Orographic enhancement brings more than 4,000 mm of annual rainfall in some parts of the Western Highlands, while further east the climate is significantly drier. Basking in the rain shadow of the Highlands, some eastern coastal areas see less than 700 mm of rain each year. Here, spring is the driest season and an upswing in easterly winds often brings coastal fog, known as haar, at this time of year. In the west, however, summer typically brings the driest months of the year, while the wettest are in late autumn and winter. Being closest to the North Atlantic storm track, it seems self-explanatory that western Scotland is incredibly wet and windy. Storms typically arrive from the west, bringing the strongest winds and heaviest rain to the Western Highlands. However, when they arrive from another direction, it can be devastating as people are caught off-guard. This is what happened in November 2021, when Storm Arwen, a powerful extratropical cyclone, arrived from the south, bringing damaging winds and rain to eastern Scotland. Millions of trees were ripped up by winds gusting to 70–80 mph across the Moray Firth and eastern regions, and thousands of people were left without power. A maximum gust of 98 mph was reached near the Borders, in Northumberland, while mountain station Cairnwell recorded a gust of 117 mph.

Scotland also experiences plenty of snowfall. This is such a key part of living in the country that researchers investigating historical language discovered that Scots have more than 400 words for 'snow', some of which are included in this book. In the centre of the Cairngorms, snow lies for more than 60 days a year on average and until recently, patches of snow would persist in the mountains all year round. This helps sustain the five ski centres dotted across the country, although climate change is threatening their future. Meanwhile, on the west coast, the number of snow days is comparable to coastal parts of England and Wales. Elsewhere, the lowlands receive a similar number to northern England.

Ireland
From towering cliffs and dramatic mountains carpeted in green, to the rolling hills and mysterious bogs found inland, the island of Ireland is teeming with unique landscapes. Exceptionally mild for its latitude and famously rainy, Ireland's climate supports an array of biodiversity and a remarkably long growing season, sustaining vast swathes of green countryside for which the Emerald Isle earns its nickname. This is largely thanks to the Atlantic Ocean, which is the dominant influence on Ireland's climate. The North Atlantic Drift, an Atlantic

ADDITIONAL INFORMATION

current extending from the Gulf Stream, transports warm waters north and eastwards towards Ireland. Seawater off the coast of Kerry has travelled all the way from Florida in this current, bringing warmth to the coastal waters of Ireland. This has a marked influence on the island's climate, keeping it relatively mild all year round and invigorating Atlantic depressions as they travel over Ireland towards Britain, increasing the amount of rainfall they bring.

Over the year, Ireland's temperatures tend to follow what the surrounding seas are doing – at their coldest in February and warmest in August. The sea also helps to mediate any extremes of temperature, so as far as there can be warm days and cold nights, Ireland rarely sees the extremes recorded in England. The hottest day on record was set more than a hundred years ago and only a few heatwaves have come close enough to challenge it since. It happened on 26 June 1887 at the Kilkenny weather station in the southeast, barely a year after it opened. An inspection that took place two years later verified that the thermometer was in good working condition, so the record still stands. Even in the heatwave of July 2022, when England, Wales and Scotland recorded new all-time highs, the 33.0 °C reached in Dublin wasn't enough to beat the 1887 record of 33.3 °C. Meanwhile, the highest temperature recorded in Northern Ireland was much more recent, with Castlederg in County Tyrone hitting 31.3 °C on 21 July 2021.

Winters in Ireland are typically mild too, thanks to the nearby Atlantic. Only strong easterly or northeasterly winds are capable of eliminating its influence, and it was under these conditions that Ireland recorded its lowest temperatures on record. Another long-standing record, temperatures plunged to −19.1 °C at Markree, County Sligo in January 1881, while more recently, the cold December of 2010 brought a record low of −18.7 °C to Castlederg in Northern Ireland.

While you can't rely on Ireland's temperatures to be particularly extreme, its rainy nature is without a doubt. Nationally, the Republic of Ireland receives nearly 1,300 mm of rain each year, while Northern Ireland's total is slightly lower, around 1,150 mm. The west of the island is by far the wettest,

ADDITIONAL INFORMATION

especially over high ground where more than 2,000 mm of rain falls in some places within a year. Generally, eastern areas are drier, especially along the coast, where some areas receive less than 900 mm in an average year. The exception to this is the Wicklow Mountains, which are wetter. Ireland's wetness is thanks to its proximity to the North Atlantic storm track, which brings a succession of depressions and associated wet and windy weather to the island, especially in early winter. At this time of year, relatively warm seas add extra energy to the systems as they approach, typically leading to more rainfall. However, come February the influence of anticyclones over Greenland and Europe tends to introduce periods of drier weather. Average monthly rainfall continues to fall into spring, with April and May being the driest months of the year before a return of the westerlies brings an increase in cloud and rain around midsummer. Towards autumn, as the Atlantic basin reaches the peak of its hurricane season, remnants of these powerful tropical storms may become mixed in with low-pressure systems to produce very energetic storms. In these scenarios, Ireland bears the brunt of the severe weather. Ex-hurricane Debbie (see page 162) and ex-hurricane Ophelia (also known as Storm Ophelia) are two of the worst ever experienced in Ireland, the latter bringing 90 mph winds, major power cuts and the loss of three lives.

Clouds

Like much of the natural world, clouds have a way of capturing our imagination. Every day, a free performance drifts across the skies, from the perfection of a single cumulus humilis cloud to the majestic power of the cumulonimbus. Clouds can be enjoyed in so many ways, whether it's spotting interesting shapes and formations or translating their movement and form to get a clue about the weather to come. Learning the basics of cloud types and what they mean can be a great way to connect with nature.

The first step in determining what kind of clouds are in the sky is estimating their height – are they low, medium or high? This is how the World Meteorological Organization categorizes clouds, which are specified by the height of their bases rather than their tops. The levels are approximate and occasionally overlap:

Low clouds: bases 2 km or lower (0–6,500 ft).
Middle clouds: bases between 2–7 km (6,500–23,000 ft).
High clouds: bases between 5–13 km (16,500–45,000 ft).

These can be further divided into ten different types, or genera, of cloud, similar to the system of categorization for plant and animal species. Each also falls into one or more of three major groups: cumulus, stratus or cirrus. These have Latin meanings that help us to understand them: cumulus means 'heap', stratus is 'layer', while cirrus comes from the word 'curl'.

ADDITIONAL INFORMATION

Low Clouds
Cumulus (Cu)
These fluffy clouds are often likened to mounds of cotton wool or cauliflower and, being the most familiar cloud type, are the easiest to recognize. Also known as 'fair-weather clouds', cumulus are convective clouds which form as air near the surface is warmed, rises, cools and condenses – a common sight on a summer's day. When heating ceases at the end of the day, they dissipate, once again leaving a blank slate of clear sky. They are individual heaps of cloud, generally well separated from one another, with rounded tops and flat, darker bases often all at the same level. The cumulus humilis is the smallest within this family, but can grow into the larger mediocris or congestus as air continues to rise, and even into the notorious cumulonimbus.

Cumulus (Cu)

Cumulonimbus (Cb)

The only cloud that reaches through all three height ranges, the cumulonimbus is the most dramatic and energetic of the cumulus family. These towering thunderclouds appear as a vast mass of heavy, dense-looking cloud that normally extends high into the sky. With a base of dark, grey, threatening cloud, the bottom of a cumulonimbus is ragged in appearance, with shafts of rain clearly visible. Meanwhile, its upper portion can reach the top of the troposphere where, with nowhere else to go, the cloud spreads out, creating the distinctive 'anvil top'. These clouds are closely associated with severe weather, including torrential rain, hail, thunder and lightning, and even tornadoes.

Cumulonimbus (Cb)

ADDITIONAL INFORMATION

Stratus (St)

Stratus (St)

If the sky was a canvas, then stratus clouds would be painted by the lazy artist simply swiping their brush from side to side, creating flattened layers of grey, indistinct cloud. They are the lowest forming of all cloudtypes, though when they appear at ground level they're known as mist or fog. Unlike cumulus clouds, which are pockets of air that cool and condense, stratus is a whole layer of air that has cooled and condensed – a process which happens if it comes into contact with a layer of denser air, thus forcing it to gently rise. When the cloud is thin enough for the Sun to be seen through it, it is known as stratus translucidus, if not, then it is stratus opacus. It rarely produces more precipitation than a little drizzle or light snow. Should low stratus form at night in the summer, bringing a foggy start, it is usually burned off by the warmth from the Sun within a few hours.

Stratocumulus (Sc)
Its hybrid name giving a clue to its appearance, stratocumulus has the layering present in stratus clouds, but with more structure, such as the lumps and bumps of a cumulus cloud. Rather than the uniform grey of a stratus cloud, it generally has more variation in tones. It usually develops in one of two ways: either from cumulus clouds spreading out to form a layer, trapped underneath an inversion, or through the break-up of a layer of stratus as shallow convection takes place. Similar to stratus, it generally only produces a small amount of precipitation.

Stratocumulus (Sc)

Medium Clouds
Altostratus (As)
A smooth blanket of cloud, not known for interesting shapes or variation in colour, altostratus is simply the mid-level version of its low-level sibling, stratus. Unlike stratus, which is composed of tiny water droplets, altostratus contains a mixture of water and ice crystals. It tends to bring dull, overcast conditions, but can be quite beautiful when illuminated by the rising or setting

ADDITIONAL INFORMATION

Altostratus (As)

sun, with gentle undulations visible at its base. As its thinner form, altostratus translucidus doesn't usually bring rain, but should it thicken to altostratus opacus or even nimbostratus, rain becomes more likely. This often happens as a warm front approaches, when the gentle lifting associated with its passage helps the air to cool and condense.

Nimbostratus (Nb)

The familiar word 'stratus' tells us this cloud is distinguished by its layers, while 'nimbo' comes from 'nimbus' which means 'rain cloud'. Put them together and you've got the higher, rain-bearing nimbostratus cloud. In colour, it appears a darker grey than the simple altostratus cloud, with shafts of precipitation often visible at its base, giving it a ragged appearance. These appendages are known as pannus. Rather than the short, sharp bursts of precipitation from a cumulonimbus cloud, nimbostratus brings more persistent rain. It doesn't grow as tall as the formidable cumulonimbus either, but can become extensive through the atmosphere, reaching from 600 metres to 7 km (2,000 to 23,000 ft).

Nimbostratus (Nb)

Altocumulus (Ac)
Often compared to bread rolls, pancakes or simply given the affectionate term 'cloudlets', altocumulus clouds have the same fluffy, clumping characteristic as cumulus clouds but occur higher up, either in localized patches or extensive layers. They often form when gentle convection is present in a layer of altostratus, breaking it up so patches of blue sky may be glimpsed in between the cloud heaps. To distinguish them from the higher cirrocumulus, altocumulus tend to display a greater range in tone, with some darker shades of grey. They are one of the most varied cloud types, with the UFO-like lenticularis and majestic castellanus also falling into this category. Lenticularis are orographic clouds that develop a distinctive lens shape as air interacts with mountain ranges, and have historically been mistaken for alien spaceships. Meanwhile castellanus are altocumulus clouds that grow small towers, giving them a castle-like appearance. These clouds indicate instability in the mid-level of the troposphere.

ADDITIONAL INFORMATION

Altocumulus (Ac)

High Clouds
Cirrus (Ci)
Another cloud that appears as if an artist took their brush to the skies, creating whimsical white streaks across a bright blue backdrop, cirrus is the highest of the common clouds. Composed entirely of ice crystals, their gently falling motion combined with high wind speeds in the upper troposphere create beautiful streaks of cloud. They are usually formed when parcels of relatively dry air move upwards, the small amount of water vapour present turning immediately to ice crystals in a process called sublimation. They often appear alone in the sky, indicating fine weather, but should they continue to thicken and advance, change is on the way. This often marks the approach of a warm front, which pushes moist air up and over the wedge of cooler air as they meet, forming these high clouds. Although these clouds do precipitate, it never reaches the

ground and instead re-evaporates. Cirrus can even form from the vapour trails left by airplanes, known as contrails.

Cirrus (Ci)

ADDITIONAL INFORMATION

Cirrostratus (Cs)
A layer of thin, white cloud high in the sky, cirrostratus adds a hazy quality to the day's sunshine and can even bring a subtle drop in temperature. Another precursor of an approaching warm front, a layer of cirrostratus tends to follow the tufts of cirrus clouds that were first to appear in the sky. It is also composed of ice crystals and formed in the same way – by gentle uplift and sublimation. Appearing as a smooth sheet or more fibrous and distinguishable strands, it is often accompanied by optical phenomena, such as sun dogs or halos (see page 90).

Cirrostratus (Cs)

Cirrocumulus (Cc)

Scattered across the skies like tiny pieces of cotton wool, cirrocumulus are perhaps the most elusive of the high clouds as they are so fleeting. Unlike cirrus and cirrostratus, these clouds contain tiny, supercooled water droplets along with the ice crystals. These supercooled particles are water molecules that stay liquid, even when below 0 °C. Cirrocumulus are formed when turbulent vertical currents meet a layer of cirrus or cirrostratus, breaking them up into smaller cloudlets. Stronger convection results in the fluffy, cotton wool-like cirrocumulus, known as floccus, while cirrocumulus stratiformis appears as a more ripple-like display. This is often referred to as 'mackerel skies' due to its resemblance to fish scales – a term that is also applied to high altocumulus. Due to the convective nature under which these clouds are formed, they tend to be rather short-lived.

Cirrocumulus (Cc)

ADDITIONAL INFORMATION

Special Clouds
In addition to the main cloud types, there are also accessory and supplementary clouds, which are features that are attached or partly merged within a cloud. Incus, referring to the anvil top of a cumulonimbus cloud; virga, the falling streaks of ice crystals from a cirrus cloud; and pannus, as mentioned earlier, are just a few examples of these.

Kelvin-Helmholtz waves
Also known as fluctus, these clouds are extremely rare and breathtaking to behold. They resemble ocean waves and are caused by strong horizontal winds along the boundary between two different layers of air. When the upper layer moves faster than the lower layer, it 'pulls' some of the existing cloud layer with it, creating the appearance of breaking waves. They can occur at any height, but are usually a higher-level cloud. This phenomenon is named after two physicists, William Thomson (Lord Kelvin) and Hermann von Helmholtz.

Kelvin-Helmholtz waves

Mammatus

Most commonly associated with cumulonimbus clouds, mammatus, also known as mamma, are smooth, udder-like protrusions emerging from the base of a cloud. Derived from the Latin for 'breast', they are caused by powerful downdrafts within the cloud from which they formed. Mammatus can vary in appearance, from near-spherical pouches to elongated tubes.

Mammatus

Asperitas

The newest cloud type, asperitas, was only added to the World Meteorological Organization's Cloud Atlas in 2017 following a proposal by the Cloud Appreciation Society in 2008. Members of the society submitted hundreds of photos of the dramatic formation, no doubt leading to its acceptance as the first new cloud type in more than 50 years. Loosely translated from the Latin for 'roughness', asperitas is a chaotic wave-like formation resembling choppy seas. Formed in unstable conditions, these clouds are often spotted near thunderstorms.

ADDITIONAL INFORMATION

Asperitas

How to Use Weather Apps

With rain one minute and sunshine the next, the weather in the UK is extremely variable. This can make planning outdoor activities a challenge, as no one wants to finish setting up their summer picnic just for the skies to open and give everything a good soaking. Meanwhile, anyone whose job involves working outside will tell you the importance of an accurate weather forecast. Industries such as agriculture, construction and transportation, to name a few, can be very sensitive to the weather. Throughout the year, farmers need to know what the weather is doing so they can make decisions about planting, spraying their crops and harvesting. A particularly accurate forecast is needed for making hay, as once it's been cut it needs to stay dry for several days before it can be baled. The importance of a good forecast cannot be understated. Luckily, there is a wealth of information out there – so much so, that it can be hard to know where to start. Using a weather app is one of the easiest ways to view a forecast. Even then, there are a plethora of different apps, each getting their data from different sources and displaying them in different ways. The simplest shows the forecast for the next few days, with complicated weather information streamlined into a series of weather symbols we all know and love. Meanwhile others include a variety of meteorological data and forecasts for more than ten days in advance. A lot of weather app forecasts come straight out of computer models, whereas some will have been tweaked by a meteorologist.

Weather symbols and the chance of rain
Weather symbols differ from one app to another, but it's generally easy to work out what they mean. Each represents the most likely weather for the day in question. For example, sunshine and showers would be shown by a sun partly obscured by a cloud, with a raindrop indicating the chance of a shower. Often below this, you'd see a percentage chance of rain.

This is a little more complicated and is where a lot of people get caught out. If there is a 90 per cent chance of rain at a given location over an hour, then that means the chance of precipitation falling at some point in that hour is nine in ten. Sometimes this is calculated by ensemble forecasting, when the same weather model is run lots of times with slightly different starting conditions. For the above example, if a model is run ten times and nine out of those ten results predict rain, then there is a 90 per cent chance of rain. When showers are expected, rather than persistent frontal rain, it becomes a bit trickier. Forecasters often say that predicting where showers are going to develop is like trying to estimate where bubbles will form in a pan of boiling water. On days like this, a weather app based on only one computer model may give one place a 90 per cent chance of rain, while another receives a 30 per cent chance, depending on where showers develop in that particular model. In reality, those showers may bubble up anywhere, and the percentage chance is more likely to be somewhere between the two. Confusing, right? But there are tools to help! Some apps also include a text forecast which might provide a clue as to whether to expect showers or rain, and a few even include a brief forecast written by a meteorologist to help you plan for your day.

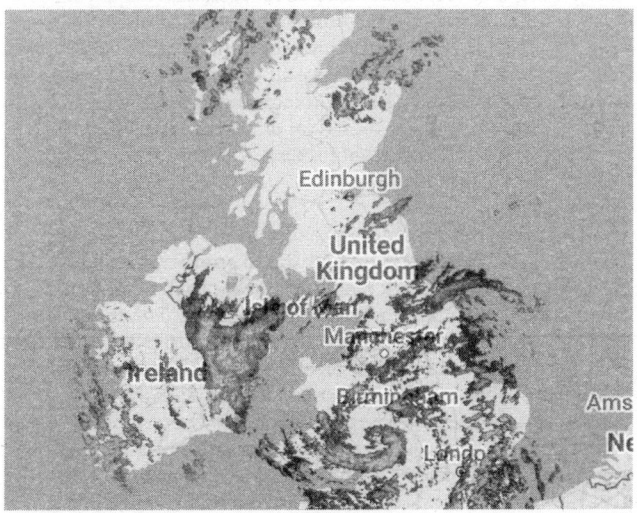

Areas of rain and showers shown by rainfall radar.

WEATHER ALMANAC 2026

Rainfall radar
Another useful tool is a rainfall radar – a map showing where precipitation is falling and its intensity. The clue is in the name: it uses radar technology to determine this. A radar emits a pulse of radio waves, which are reflected back towards the radar when they hit rain. Based on the strength of these echoes and the time it takes for them to return, the position and intensity of the rain can be calculated. On some apps, this is updated every five minutes, allowing you to see where it's raining nearby and whether it's moving towards or away from you. There may even be a map which shows the expected distribution of rainfall over the next hour or the coming days.

Nowcasting
You might notice that the forecasts in weather apps change a lot, particularly during periods of unsettled weather. They are constantly being updated with new observations and the latest model runs. Some even use a process called 'nowcasting', which updates predictions for the next few hours based on what's happening in the present.

ADDITIONAL INFORMATION

The Beaufort Scale

Wind strength is commonly given on the Beaufort scale. This was originally defined by Francis Beaufort (later Admiral Beaufort) for use at sea, but was subsequently modified for use on land. Meteorologists generally specify the speed of the wind in metres per second (m s^{-1}). For wind speeds at sea, details are usually given in knots.

The Beaufort scale (for use at sea)

Force	Description	Sea state	Speed (Knots)	Speed (ms^{-1})
0	calm	like a mirror	<1	0.0–0.2
1	light air	ripples, no foam	1–3	0.3–1.5
2	light breeze	small wavelets, smooth crests	4–6	1.6–3.3
3	gentle breeze	large wavelets, some crests break, a few white horses	7–10	3.4–5.4
4	moderate breeze	small waves, frequent white horses	1–16	5.5–7.9
5	fresh breeze	moderate, fairly long waves, many white horses, some spray	17–21	8.0–10.7
6	strong breeze	some large waves, extensive white foaming crests, some spray	21–27	10.8–13.8
7	near gale	sea heaping up, streaks of foam blowing in the wind	28–33	13.9–17.1
8	gale	fairly long and high waves, crests breaking into spindrift, foam in prominent streaks	34–40	17.2–20.7

The Beaufort scale (for use at sea) – *continued*

9	strong gale	high waves, dense foam in wind, wave crests topple and roll over, spray interferes with visibility	41–47	20.8–24.4
10	storm	very high waves with overhanging crests, dense blowing foam, sea appears white, heavy tumbling sea, poor visibility	48–55	24.5–28.4
11	violent storm	exceptionally high waves may hide small ships, sea covered in long, white patches of foam, waves blown into froth, poor visibility	56–63	28.5–32.6
12	hurricane	air filled with foam and spray, visibility extremely bad	64	32.7

The Beaufort scale (adapted for use on land)

Force	Description	Events on land	Speed	
			Knots	ms^{-1}
0	calm	smoke rises vertically	<1	0.0–0.2
1	light air	direction of wind shown by smoke but not by wind vane	1–3	0.3–1.5
2	light breeze	wind felt on face, leaves rustle, wind vane turns to wind	4–6	1.6–3.3

ADDITIONAL INFORMATION

The Beaufort scale (adapted for use on land) – *continued*

3	gentle breeze	leaves and small twigs in motion, wind spreads small flags	7–10	3.4–5.4
4	moderate breeze	wind raises dust and loose paper, small branches move	11–16	5.5–7.9
5	fresh breeze	small leafy trees start to sway, wavelets with crests on inland water	17–21	8.0–10.7
6	strong breeze	large branches in motion, whistling in telephone wires, difficult to use umbrellas	21–27	10.8–13.8
7	near gale	whole trees in motion, difficult to walk against wind	28–33	13.9–17.1
8	gale	twigs break from trees, difficult to walk	34–40	17.2–20.7
9	strong gale	slight structural damage to buildings; chimney pots, tiles and aerials removed	41–47	20.8–24.4
10	storm	trees uprooted, considerable damage to buildings	48–55	24.5–28.4
11	violent storm	widespread damage to all types of buildings	56–63	28.5–32.6
12	hurricane	widespread destruction, only specially constructed buildings survive	64	32.7

The International Tornado Intensity Scale

The TORRO tornado intensity scale is based on an extension to the Beaufort scale of wind speeds. The winds speeds are actually calculated mathematically from the accepted Beaufort wind speeds. (Although the Beaufort scale was first proposed in 1805, it was expressed in terms of wind speed in 1921.) T0 corresponds to Beaufort Force 8, and T11 would correspond to Beaufort Force 30 (if such a force existed).

The TORRO scale is thus solely based on wind speeds, unlike the Fujita scale and the later, modified version, the Enhanced Fujita scale, which are based on an assessment of damage. In practice, wind-speed measurements are rarely available for tornadoes, and so, in effect, both scales are, perforce, based on an assessment of the intensity of damage.

Scale	Wind speed (estimated)			Potential damage
	mph	kmh^{-1}	ms^{-1}	
F0	0–38	0–60	0–16	**No damage.** *(Funnel cloud aloft, not a tornado)* No damage to structures, unless on tops of tallest towers, or to radiosondes, balloons and aircraft. No damage in the country, except possibly agitation to highest tree-tops and effect on birds and smoke. A whistling or rushing sound aloft may be noticed.
T0	39–54	61–86	17–24	**Light damage.** Loose light litter raised from ground-level in spirals. Tents, marquees seriously disturbed; most exposed tiles, slates on roofs dislodged. Twigs snapped; trail visible through crops.

ADDITIONAL INFORMATION

T1	55–72	87–115	25–32	**Mild damage.** Deckchairs, small plants, heavy litter becomes airborne; minor damage to sheds. More serious dislodging of tiles, slates, chimney pots. Wooden fences flattened. Slight damage to hedges and trees.
T2	73–92	116–147	33–42	**Moderate damage.** Heavy mobile homes displaced, light caravans blown over, garden sheds destroyed, garage roofs torn away. Much damage to tiled roofs and chimney stacks. General damage to trees, some big branches twisted or snapped off, small trees uprooted.
T3	93–114	148–184	42–51	**Strong damage.** Mobile homes overturned/badly damaged; light caravans destroyed; garages and weak outbuildings destroyed; house roof timbers considerably exposed. Some larger trees snapped or uprooted.

T4	115–136	185–220	52–61	**Severe damage.** Motor cars levitated. Mobile homes airborne / destroyed; sheds airborne for considerable distances; entire roofs removed from some houses; roof timbers of stronger brick or stone houses completely exposed; gable ends torn away. Numerous trees uprooted or snapped.
T5	137–160	221–259	62–72	**Intense damage.** Heavy motor vehicles levitated; more serious building damage than for T4, yet house walls usually remaining; the oldest, weakest buildings may collapse completely.
T6	161–186	260–299	73–83	**Moderately devastating damage.** Strongly built houses lose entire roofs and perhaps also a wall; windows broken on skyscrapers, more of the less-strong buildings collapse.
T7	187–212	300–342	84–95	**Strongly devastating damage.** Wooden-frame houses wholly demolished; some walls of stone or brick houses beaten down or collapse; skyscrapers twisted; steel-framed warehouse-type constructions may buckle slightly. Locomotives thrown over. Noticeable debarking of trees by flying debris.

ADDITIONAL INFORMATION

T8	213–240	343–385	96–107	**Severely devastating damage.** Motor cars hurled great distances. Wooden-framed houses and their contents dispersed over long distances; stone or brick houses irreparably damaged; skyscrapers badly twisted and may show a visible lean to one side; shallowly anchored high rises may be toppled; other steel-framed buildings buckled.
T9	241–269	386–432	108–120	**Intensely devastating damage.** Many steel-framed buildings badly damaged; skyscrapers toppled; locomotives or trains hurled some distances. Complete debarking of any standing tree trunks.
T10	270–299	433–482	121–134	**Super damage.** Entire frame houses and similar buildings lifted bodily or completely from foundations and carried a large distance to disintegrate. Steel-reinforced concrete buildings may be severely damaged or almost obliterated.

T11	>300	>483	>135	**Phenomenal damage.** Strong framed, well-built houses leveled off foundations and swept away. Steel-reinforced concrete structures are completely destroyed. Tall buildings collapse. Some cars, trucks and train carriages may be thrown approximately 1 mile (1.6 kilometres).

The TORRO Hailstorm Intensity Scale

The Tornado and Storm Research Organisation (TORRO) has not only developed a scale for rating tornadoes (see pages 252–256) but also one to judge the severity of hailstorm incidents. This scale is given in the following table, but it must be borne in mind that the severity of any hailstorm will depend (among other factors) upon the size of individual hailstones, their numbers and also the speed at which the storm itself travels across country.

Scale	Intensity	Hail size (mm)	Size comparison	Damage
H0	Hard hail	5–9	Pea	None
H1	Potentially damaging	10–15	Mothball	Slight general damage to plants, crops
H2	Significant	16–20	Marble, grape	Significant damage to fruit, crops, vegetation
H3	Severe	21–30	Walnut	Severe damage to fruit and crops; damage to glass and plastic structures; paint and wood scored
H4	Severe	31–40	Pigeon's egg > squash ball	Widespread damage to glass; damage to vehicle bodywork
H5	Destructive	41–50	Golf ball > pullet's egg	Wholesale destruction of glass; damage to tiled roofs; significant risk of injuries

H6	Destructive	51–60	Hen's egg	Bodywork of grounded aircraft dented; brick walls pitted
H7	Destructive	61–75	Tennis ball > cricket ball	Severe roof damage; risk of serious injuries
H8	Destructive	76–90	Large orange > softball	Severe damage to aircraft bodywork
H9	Super hailstorms	91–100	Grapefruit	Extensive structural damage
H10	Super hailstorms	>100	Melon	Extensive structural damage; risk of severe or fatal injuries to persons caught in the open

ADDITIONAL INFORMATION

Twilight Diagrams

For each individual month, we give details of sunrise and sunset times for the four capital cities of the various countries that make up the United Kingdom.

During the summer, especially at high latitudes, twilight may persist throughout the night and make it difficult to see the faintest stars. Beyond the Arctic and Antarctic Circles, of course, the Sun does not set for 24 hours at least once during the summer (and rise for 24 hours at least once during the winter). Even when the Sun does dip below the horizon at high latitudes, bright twilight persists throughout the night, so observing the fainter stars is impossible. Even in Britain this applies to northern Scotland, which is why we include a diagram for Lerwick in the Shetland Islands.

As mentioned earlier (page 9) there are three recognised stages of twilight: civil twilight, nautical twilight and astronomical twilight. Full darkness occurs only when the Sun is more than 18° below the horizon. During nautical twilight, only the very brightest stars are visible. During astronomical twilight, the faintest stars visible to the naked eye may be seen directly overhead, but are lost at lower altitudes. They become visible only once it is fully dark. The diagrams show the duration of twilight at the various locations. Of the locations shown, during the summer months there is astronomical twilight for a short time at Belfast, and this lasts longer during the summer at all of the other locations. To illustrate the way in which twilight occurs in the far south of Britain, we include a diagram showing twilight duration at St Mary's in the Scilly Isles. (A similar situation applies to the Channel Islands, which are also in the far south.) Once again, full darkness never occurs.

To get a fuller idea of the degree of darkness on a given night, the diagrams should be used in conjunction with the monthly moon diagrams, and the moonrise and moonset tables. The lunar phase, and whether the moon is above the horizon, significantly affects general visibility.

WEATHER ALMANAC 2026

Belfast, UK – Latitude 54.6°N – Longitude 5.8°W

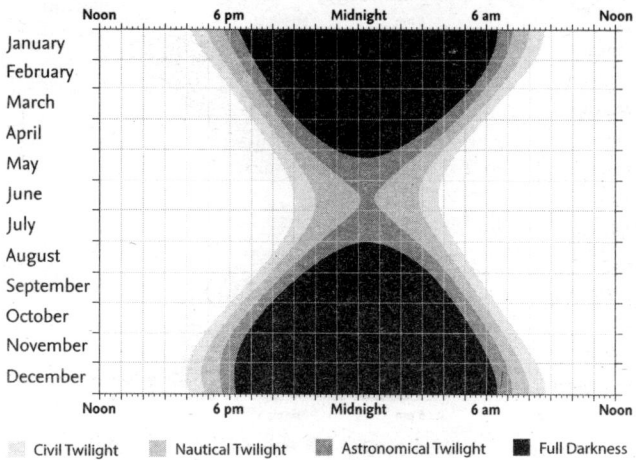

Cardiff, UK – Latitude 51.5°N – Longitude 3.2°W

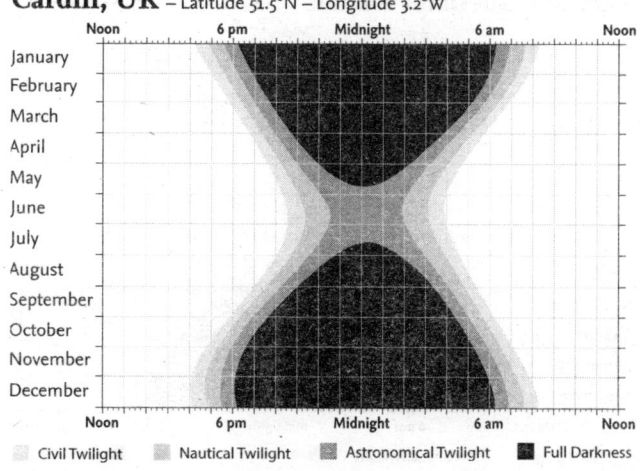

ADDITIONAL INFORMATION

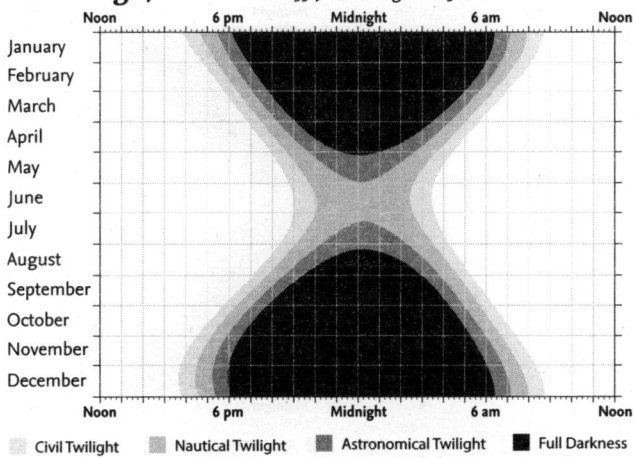

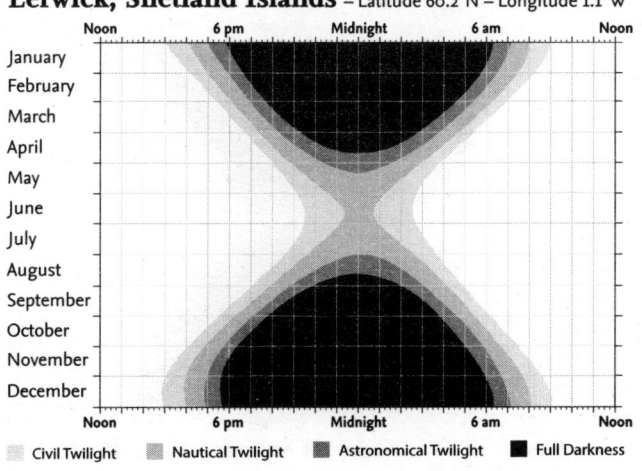

WEATHER ALMANAC 2026

London, UK – Latitude 51.5°N – Longitude 2.0°W

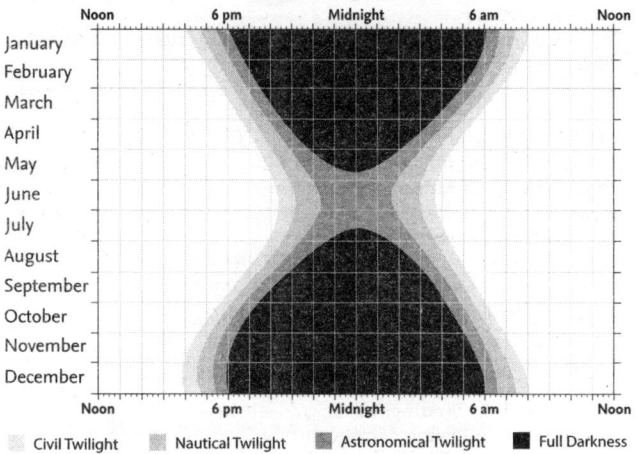

St Mary's, Scilly Isles – Latitude 49.9°N – Longitude 6.4°W

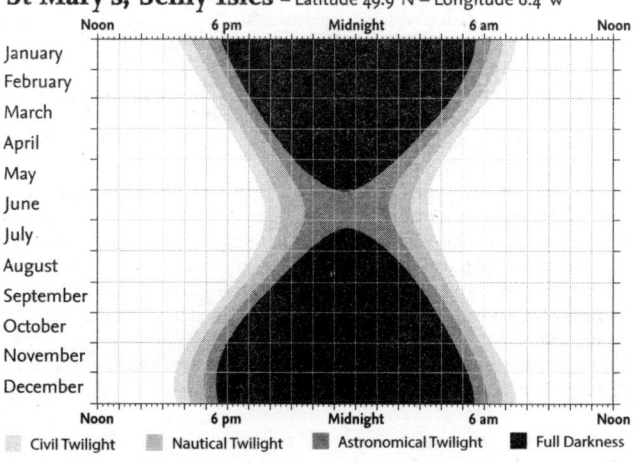

References

Brooks, Norman. 'The August 1912 floods in Norfolk.' *Weather* 67, no. 8 (2012): 204-205.

Coonan, Barry, Mary Curley, and Ciara Ryan. 'Long-term Rainfall Averages for Ireland 1991-2020.' (2024).

Dictionary of the Scots Language. 2004. Scottish Language Dictionaries Ltd.

Doe, Robert K., and Paul R. Brown. 'A Sea On The Moors: Devastating Flash Flooding At Holmfirth, West Yorkshire, 29 May 1944.' *Journal Of Meteorology* 30, no. 298 (2005): 163-175

Goose, Herbert H. *Norwich Under Water: Reminiscences of The Two Great Floods*. Goose & Son.

Grahame, Nick, Bob Riddaway, Alison Eadie, Baden Hall, and Ewen McCallum. 'Exceptional hailstorm hits Ottery St Mary on 30 October 2008.' *Weather* 64, no. 10 (2009): 255-263.

Harrison, R. Giles, and Stephen D. Burt. 'Accuracy of daily extreme air temperatures under natural variations in thermometer screen ventilation.' *Atmospheric Science Letters* 25, no. 10 (2024): e1256.

Horton, Sarah L. 'Findings of tornado site investigations undertaken following damage during Storm Ciarán on 1–2 November 2023.' *Weather* 79, no. 11 (2024): 373-378.

Kendon, M. *Storm Ciarán, 1 to 2 November 2023*. 2023.

Kendon, Mike, and Mark McCarthy. 'The UK's wet and stormy winter of 2013/2014.' *Weather* 70, no. 2 (2015): 40-47

Kew, Sarah F., M. McCarthy, C. Ryan et al. 'Autumn and Winter Storms over UK and Ireland Are Becoming Wetter Due to Climate Change.' Grantham Institute for Climate Change, (2024).

Knightley, Paul, Sarah Horton, Matthew Clark, and Matthew Winter. 'The Jersey tornado and hailstorm of 1–2 November 2023.' *Weather* 79, no. 3 (2024): 81-84.

Lamb, Hubert H. 'Types and Spells of Weather around the Year in the British Isles : Annual Trends, Seasonal Structure of the Year, Singularities.' *Quarterly Journal of the Royal Meteorological Society* 76, Issue 330 (1950): 393-429.

Lapworth, Alan, and James McGregor. 'Seasonal Variation of the Prevailing Wind Direction in Britain.' *Weather* 63, Issue 12 (2008): 365-368.

Mayes, Julian et al. 'Regional weather and climates of the British Isles.' 9-part series. *Weather* 68-69 (2013-14)

Met Office. https://www.metoffice.gov.uk/.

Pedgley, D. E. 'A remarkable winter cold front hailstorm, 22 February 1995.' *Weather* 51, no. 10 (1996): 330-341.

Photographs of the Floods in Norwich + Norfolk August 1912. A. E. Coe & Son.

Pope, R. J., A. M. Marshall, and B. O. O'Kane. 'Observing UK Bonfire Night pollution from space: analysis of atmospheric aerosol.' *Weather* 71, no. 11 (2016): 288-291.

Simmonds, J. 'The English tornadoes of 21 May 1950.' Weather 5, no. 7 (1950): 255-257.

Storey, Neil R. *Norfolk Floods: An Illustrated History of 1912, 1938 & 1953*. Halsgrove, 2012.

Stott, Peter A., Dáithí A. Stone, and Myles R. Allen. 'Human Contribution to the European Heatwave of 2003.' *Nature* 432, (2004): 610–614.

Time and Date. https://www.timeanddate.com/.

Zachariah, Mariam, Robert Vautard, Dominik L. Schumacher et al. 'Without human-caused climate change temperatures of 40 C in the UK would have been extremely unlikely.' (2022).

Image Sources

- p. 7 Julia Murray
- p. 13 NOAA Global Monitoring Laboratory
- p. 15 Met Office
- p. 26 Zoë Johnson
- p. 30 NASA Earth/ZUMA Press Wire/Shutterstock
- p. 31 Freie Universität Berlin
- p. 42 James Bell/Alamy Stock Photo
- p. 47 Met Office
- p. 63 DPA Picture Alliance/Alamy Stock Photo
- p. 73 Zoë Johnson
- p. 74 Met Office
- p. 79 Jeff Morgan 16/Alamy Stock Photo
- p. 90 Zoë Johnson
- p. 95 Pictorial Press Ltd/Alamy Stock Photo
- p. 106 Shutterstock
- p. 111 William Edwards/Getty Images
- p. 111 Wade Sheldon Photography
- p. 121 Zoë Johnson
- p. 122 Zoë Johnson
- p. 127 Shutterstock
- p. 139 Zoë Johnson
- p. 144 Chronicle/Alamy Stock Photo
- p. 154 National Maritime Museum
- p. 159 Smith Archive/Alamy Stock Photo
- p. 171 Zoë Johnson
- p. 176 Met Office
- p. 185 Shutterstock
- p. 186 Shutterstock
- p. 191 Stringer/Getty Images
- p. 203 Axon/Alamy Stock Photo
- p. 208 Bertrand Guay/Getty Images
- p. 235 Zoë Johnson
- p. 236 Shutterstock
- p. 237 Zoë Johnson
- p. 238 Shutterstock
- p. 239 Shutterstock
- p. 240 Shutterstock
- p. 241 Zoë Johnson
- p. 242 Zoë Johnson
- p. 243 Shutterstock
- p. 244 Zoë Johnson
- p. 245 Shutterstock
- p. 246 Shutterstock
- p. 247 Shutterstock
- p. 249 WQ Radar

Acknowledgements

Thank you very much to the team at Collins: to my editor Sam Fitzgerald, for your valuable guidance, suggestions and never-ending patience answering all my questions, to Gerry Breslin for giving me the chance to write this book and to Dr Ryan French, solar physicist and author, for the recommendation that started it all.

Thanks also to Julia Murray for her beautiful cover and illustrations.

Thanks to my mum, Sarah Johnson, for the hours spent reading everything I've ever written, from university dissertations to my first published book. Agonising over every little detail is more fun with you by my side!

Thanks to Aisling Creevey for the years of kindness and support you've shown me and for giving your time and meteorological experience as a proof-reader.

To my husband Tom, my friends and family for being the best cheerleaders I could ask for, and to fellow meteorologists and weather enthusiasts Richard Jones, Fred Best and Chris Page for helping fact-check and offering advice.

And finally, thanks to the readers, I hope you enjoy the Weather Almanac 2026.

This book is dedicated to Storm Dunlop FRMetS (1942-2025). Storm leaves behind an incredible legacy, from his many contributions to the meteorological world to his work on previous editions of the Weather Almanac. Storm's passion for the weather inspired many and he will be sorely missed.

Index

AI4NWP 175
air frost 23, 39, 55, 71, 87, 103, 135, 151, 167, 183, 199, 216
Alan Turing Institute 175
anticyclones 25, 98, 119, 130, 146, 182, 199
anticyclonic gloom 182, 183
aphelion 121
April 65–79, 92
Arctic 4, 18, 22, 30, 51, 102, 150, 183
Arctic sea ice 195
artificial intelligence (AI) 175–6
asteroid impacts 195
Atlantic 4, 18, 32, 39, 50, 57, 98, 119, 158, 166–7, 178, 199, 211, 214, 221, 224, 227, 230–3
atmospheric blocking 6
'atmospheric river' 221
attribution studies 14
August 129–44
aurora borealis 86, 151
autumn 6, 7, 10, 30, 94, 150–1, 162, 169, 185, 214, 219, 222–6, 230, 233
astronomical 146, 153
meteorological 146, 152, 153
autumn equinox 6, 146, 153, 156, 157
autumnal 8
avalanches 200

'Beast from the East' 32, 42, 216
Beaufort Scale 154, 251–3, 254
Belfast
moonrise/moonset 27, 43, 59, 75, 91, 107, 123, 140, 155, 172, 187, 204
sunrise/sunset 8, 27, 43, 59, 75, 91, 107, 123, 140, 155, 172, 187, 204
twilight diagrams 28, 44, 60, 76, 92, 108, 124, 141, 156, 173, 188, 205, 262
Berlin Mandate 62
blocking highs 57, 82
Bonfire (Guy Fawkes) Night 178, 188
British Double Summer Time 94
British Summer Time (BST) 8, 57, 94–5
Brown Willy effect 131
bullet staines 179

carbon dioxide 12, 13, 56, 207
Cardiff
moonrise/moonset 27, 43, 59, 75, 91, 107, 123, 140, 155, 172, 187, 204
sunrise/sunset 8, 27, 43, 59, 75, 91, 107, 123, 140, 155, 172, 187, 204
twilight diagrams 28, 44, 60, 76, 92, 108, 124, 141, 156, 173, 188, 205, 262
Celsius 168
Channel Islands 211–14
Chernobyl disaster 78–9
Christmas 194, 195, 199
climate, global 12–15
climate change 12–15, 62–3, 110–11, 130, 216–17
human-induced 12, 13, 114
climate warming stripes 110–11
climatic regions 211–33
cloud in a bottle 186

cloud condensation nuclei 178, 186
clouds 162, 182, 207, 234–47
altocumulus (Ac) 240–1
altostratus (As) 74, 238–9
asperitas 246–7
castellanus 240
cirrocumulus (Cc) 244
cirrostratus (Cs) 90, 243
cirrus (Ci) 74, 90, 234, 241–2, 245
cloud cover 58
convective 34, 235
cumulonimbus (Cb) 74, 82, 104, 234–6, 245, 246
cumulus (Cu) 34, 74, 234, 235, 237–8
cumulus humilis 235
floccus 244
and frontal systems 74
funnel 83
Helm Bar 66
high 234, 241–4
Kelvin-Helmholtz waves (fluctus) 245
lenticularis 202, 240
low 234, 235–8
mammatus 246
medium 234, 238–40
nacreous 202–3
nimbostratus (Nb) 74, 239–40
noctilucent 122, 202
pannus 239, 245
special 245–7
stratocumulus (Sc) 238
stratus (St) 234, 237–8
contrails 242
Coordinated Universal Time (UTC) 8
Copenhagen Accord 62
COPs 62–3, 200

INDEX

crepuscular rays 137
cyclones
  extratropical 41, 70, 99, 153
  tropical 158
  see also anticyclones

Daylight Saving Time (DST) 8, 94–5
December 178, 193–208
depressions (low-pressure systems) 4, 18, 24, 34, 38–9, 40, 50, 54–5, 57, 70–1, 87–9, 98–9, 102–3, 118, 134–5, 143, 153, 162, 166, 168, 178, 194, 198–9, 214, 221, 224, 233
diamond dust 201
dinderex 131
'dog days' 114–15
dreich 35
drookit (drook) 163
droughts 14, 216–17
dusk *see* twilight
dust devils 105

Earth's tilt 6, 8
East Anglia 214–17
eclipses
  partial lunar 137, 142
  solar 136–9
Edinburgh
  moonrise/moonset 27, 43, 59, 75, 91, 107, 123, 140, 155, 172, 187, 204
  sunrise/sunset 8, 27, 43, 59, 75, 91, 107, 123, 140, 155, 172, 187, 204
  twilight diagrams 28, 44, 60, 76, 92, 108, 124, 141, 156, 173, 188, 205, 263
El Niño 12
El Niño Southern Oscillation (ENSO) 179, 194

equilux 156
equinoxes
  autumn 6, 146, 153, 156, 157
  spring 6, 57, 60
estival 8
European Centre for Medium-Range Weather Forecasts (ECMWF) 176, 207
ex-hurricane Debbie 158–9, 233
ex-hurricane Kirk 166
ex-hurricane Ophelia 233
ex-hurricanes 158–9
'explosive cyclogenesis' ('weather bomb') 153, 190

Fahrenheit 168
FastNet machine-learning weather model 175
February 33–47
feefle 19
fierce mild 51
flaggie 195
flooding 10, 12, 14, 22, 24, 34, 38–40, 47, 55–6, 71, 83, 86–8, 102, 110, 120, 131, 134, 136, 150–1, 166, 184, 214, 217
  Great Flood, 1912 143–4
Foehn effect 23, 25, 182, 226, 229
fog 168–9, 178, 199, 213, 224, 237
  advection 82, 89
  radiation 185
fogbows 41
fossil fuels 12–13
fret 83, 224
frost 86, 146, 150, 178, 216, 217
  see also air frost
frozen bubbles 26

gales 88, 143, 152, 169, 194, 200, 252–3
  equinoctial 146
gloaming (gloamin) 147
global warming 13, 62, 110–11
Great Blizzard, 1891 56
Great Northamptonshire Hailstorm 153
Great Sheffield Gale 40
Great Storm of 1703 184
Great Storm of 1987 190
greenhouse effect 13
greenhouse gases 12–13, 14–15, 114
Greenwich Mean Time (GMT) 8, 94

hail 24, 35, 54, 82–3, 86, 104, 120, 121, 136, 152–3, 163, 178–9, 190, 200, 260
Hailstorm Intensity Scale 259–60
halo 243
heatwaves 14–15, 104, 110, 114, 130, 134, 152, 226, 232
Helm Wind 66
hibernal 8
high-pressure systems 22, 24, 25, 38, 54, 57, 78, 86, 102–3, 119, 134, 151, 166, 182, 183, 194, 198–9
historic events 24–5, 40–1, 56, 72, 88–9, 104–5, 120, 136–7, 152–3, 168–9, 184, 200
humidity 134
hurricanes 12, 135, 158–9, 233, 252–3

ice 26, 40, 90, 122, 184, 201, 241, 243, 244, 245
ice cores 13
incus 245
Intergovernmental Panel on Climate Change 13

International Tornado Intensity Scale 254–8
Ireland 230–3
Isle of Man 219–22

January 17–32, 34
jet stream 4, 6, 32, 34, 98–9, 118, 134, 162
July 113–27, 135
June 97–111, 135
'13 June enigma' 104

Keeling, Charles 56
Keeling Curve 56
Köppen climate classification 211
Kyoto Protocol 62

La Niña 179, 194
Lamb, Hubert 7
land-based convection systems 34, 162, 213
latitude 8
Lerwick, Shetland Islands, twilight diagram 261, 263
lightning 99, 121, 131, 200, 216, 236
ball 168
line convection 35, 169
London
  moonrise/moonset 27, 43, 59, 75, 91, 107, 123, 140, 155, 172, 187, 204
  sunrise/sunset 8, 27, 43, 59, 75, 91, 107, 123, 140, 155, 172, 187, 204
  twilight diagrams 28, 44, 60, 76, 92, 108, 124, 141, 156, 173, 188, 205, 264
low-pressure systems *see* depressions
lunar eclipses, partial 137, 142

March 49–63
May 81–95

meridional 4
mesocyclones 121
Met Éireann 147, 166–7
Met Office 5, 12, 14–15, 19, 23, 50, 74, 79, 104, 115, 130, 135, 147, 154, 175–6, 184, 190, 198, 207–8
Met Office Hadley Centre 14
Met Office HadUK data set 5
methane 13
Midlands 217–19
mist 178, 199
Moon
  Beaver 189
  Buck 125
  by month 25, 29, 41, 45, 57, 61, 73, 77, 89, 93, 105, 109, 121, 125, 137, 142, 151, 153, 157, 169, 174, 185, 189, 201, 206
  Cold 206
  Flower 93
  Full 8–9, 29, 45, 61, 77, 93, 109, 125, 142, 157, 174, 206
  Full Corn 153, 157
  Harvest 151, 153, 157
  Hunter's 174
  New 8–9
  orbit 8
  Pink 77
  Snow 45
  Strawberry 109
  Sturgeon 142
  Wolf 29
  Worm 61
Moon phases 8–9, 29, 45, 61, 77, 93, 109, 125, 157, 174, 189, 206
moonrise 27, 43, 59, 75, 91, 107, 123, 140, 155, 172, 187, 204
moonset 27, 43, 59, 75, 91, 107, 123, 140, 155, 172, 187, 204

'Morning Cloud Storm' 152
'Muckle Spate' 136

Native Americans 9
nights, 'tropical' 114
North Atlantic Drift 230–2
North Atlantic Oscillation (NAO) 194
North Sea 18, 24, 46, 82, 216
Northeast England 222–4
Northwest England 219–22
November 177–91
nowcasting 250
Numerical Atmospheric-dispersion Modelling Environment (NAME) 79

observing weather 26, 42, 58, 74, 90, 106, 122, 138–9, 154, 170–1, 186
October 157, 161–76
oktas 58
orographic effects 50, 202, 213, 221, 226–7, 229, 240
ozone layer 202

Paris Agreement 12, 62–3, 200
'Pembrokeshire Dangler' 227
perihelion 25
pollen 99
pressure
  maximum/minimum 20–1, 36–7, 52–3, 84–5, 100–1, 116–17, 132–3, 148–9, 164–5, 180–1, 196–7
  see also depressions; high-pressure systems
prevernal 8

INDEX

radioactive material 78–9
radiosondes 32, 254
rain gauges 170–1
'rain shadow' 216, 219, 222, 224, 226
rainbows 73
rainfall 10–12, 14, 236, 239
  April 70–3
  August 130–1, 134–6, 143–4
  by climatic region 213–14, 216, 219, 221–2, 224, 226–7, 229–30, 232–3
  chance of 248–9
  December 194, 198–200
  February 34–5, 38–9
  freezing 18, 24
  January 18, 22–4
  July 118, 120
  June 98, 99, 102–3
  March 54–6
  May 82–3, 86–9
  November 178, 182–4
  October 162–3, 166–7, 169
  radioactive 78
  September 150–2
rainfall radar (tool) 250
Rayleigh scattering 18
refraction 90
'return of the westerlies' 98

St Mary's, Scilly Isles, twilight diagram 261, 264
Scherhag, Richard 30–2
Scotland 227–30
sea temperatures 51
seasons 6–8, 7
September 145–59
serotinal 8
severe weather 14
Siberia 32, 195, 216
Sirius 114–15
sleet 24, 54, 55, 178, 200
Slingo, Professor Dame Julia 207–8

snow 11, 46–7, 178–9, 200, 216, 222, 227, 230
  April 67, 70, 72
  December 195, 198, 199, 200, 201
  depth 42
  February 32, 38, 40, 42
  January 18–19, 22, 24
  July 115
  June 104
  March 50, 54–6
  May 88
  November 183, 184
snowstorms 19, 46–7, 56, 72, 178–9
solar eclipses 138–9
  how to view 138–9
  partial 136, 137, 138–9
  total 136, 138
solstices 6
  summer 6, 98, 105, 106, 108, 110
  winter 6, 29, 194, 199, 201, 205–6
Southeast England 214–17
Southwest England 211–14
spring 6–7, 10, 61, 66–7, 70, 77, 86–8, 94, 98, 125, 162, 213–14, 216, 219, 222, 224, 226–7, 229, 233
  astronomical 57
  meteorological 50, 57, 82
spring equinox 6, 57, 60
squall lines 35
Stevenson Screen 126–7
Storm Abigail 184
Storm Arwen 230
Storm Ashley 167
Storm Bert 10, 183
Storm Ciarán 190–1
Storm Conall 183
Storm Darragh 10, 198
Storm Eunice 10, 214
Storm Freya 50
Storm Gareth 50
Storm Henk 10, 22
Storm Isha 10, 22–3

Storm Jake 50
Storm Jocelyn 23
Storm Kathleen 70
Storm Lilian 10, 130, 135
Storm Pierrick 70–1
Storm Ulysses 41
storms 4, 10, 14, 34–5, 50, 72, 98, 104, 110, 125, 130, 136, 146–7, 153, 158, 162–3, 178–9, 184, 194, 200, 214, 222, 224, 226, 230, 233, 252–3, 260
  see also cyclones; ex-hurricanes; hurricanes; snowstorms; thunderstorms; tornadoes; typhoons
stratospheric polar vortex 30–2, 194–5, 202
sudden stratospheric warmings 30–2
summer 4, 6–8, 7, 10–11, 14, 57, 82, 94, 98–9, 104, 109–10, 114, 118–19, 121–2, 125, 130, 134–5, 141, 143, 146, 162–3, 169, 202, 213–17, 219, 221–2, 224, 226–7, 229–30, 233, 235, 237, 248, 261
  astronomical 98, 105
  meteorological 98, 104, 105
summer solstice 6, 98, 105, 106, 108, 110
Sun 6, 8, 9, 13, 18, 60, 90, 92, 137, 156, 162–3, 173, 185, 202, 237, 261
sun dogs (parhelia) 90, 243
sundials 106
sunrise 8, 27, 43, 59, 75, 91, 107, 123, 140, 155, 172, 187, 204
sunset 8, 27, 43, 59, 75, 76, 91, 107, 123, 140, 155, 172, 173, 187, 204
sunshine
  April 70–1, 73, 92

August 134, 135, 137
by climatic region 214, 226
December 194, 198, 199
February 38, 39
January 22, 23
July 118
June 98, 102–3
March 54–5
May 82, 86, 87
November 182–3
October 166–7
September 150–1
swullocking 115

temperatures 12
annual mean 11, 12
April 68–72
August 130, 132–6
by climatic region 213, 216–17, 221–2, 226–7, 229, 232
December 196–9
February 36–9
global 63, 110, 200
January 18–19, 20–3
July 114–19
June 100–4
March 50–5
maximum 11, 20–3, 36–9, 50–5, 68–71, 82, 84–7, 100–4, 114, 116–19, 130, 132–6, 148–52, 164–7, 180–3, 196–9, 217
May 82, 84–8
minimum 11, 18–19, 20–3, 36–9, 52–5, 68–72, 84–8, 100–3, 115, 116–19, 132–5, 148–51, 164–7, 180–3, 196–9
November 179, 180–3
October 164–7
sea 51
September 148–52
'tropical' nights 114
thermometers 126–7

thunderplump 99
thunderstorms 6, 34–5, 35, 82–3, 86–9, 99, 102, 120–1, 130–1, 131, 134–5, 150, 162, 168, 175, 190, 200, 214, 216, 219, 224, 227, 236, 246
supercell 120, 121
time 8, 57, 94–5, 106
Tornado and Storm Research Organisation, The (TORRO) 82
Hailstorm Intensity Scale 259–60
International Tornado Intensity Scale 254–8
tornadoes 24, 72, 83, 120, 121, 136–7, 151, 168, 200, 216, 236, 254–8
Jersey 190–1
twilight 9
astronomical 9, 108, 124, 261
civil 9, 108, 261
nautical 9, 108, 261
twilight diagrams 9, 28, 44, 60, 76, 92, 108, 124, 141, 156, 173, 188, 205, 261–4
typhoons 12

vernal 8
virga 245

Wales 224–7
water vapour 13, 78, 89, 122, 178, 186, 201, 221, 224, 241
waterspouts 24
weather apps 248–50
weather extremes 20–1, 36–7, 52–3, 68–9, 84–5, 100–1, 116–17, 132–3, 148–9, 164–5, 180–1, 196–7
weather forecasts 248–50
weather fronts, tracking 74

weather symbols 248–9
Willet, William 94, 95
wind 11, 213
April 66, 70
August 134–7, 143
by climatic region 219, 222, 226, 230, 232–3
Custard 66
December 194, 198–200
February 34, 35, 38, 40
Fen Blows 66
January 22–3, 24, 30
July 118, 120
June 102
March 54–6
May 86, 88
November 183, 184, 190–1
October 166, 169
September 150, 152, 158
speed/strength 154, 251–3, 254
TORRO Tornado Scale 254–8
winter 4, 6–8, 10, 14, 18–19, 22, 26, 29–30, 34–5, 39, 45, 50–1, 55, 61, 67, 110, 162–3, 169, 174, 179, 185, 189–90, 194–5, 198–9, 201–2, 213–14, 216–17, 219, 221–2, 224, 226–7, 229–30, 232–3, 261
1947 46–7
astronomical 194, 201
meteorological 18, 194, 201
winter solstice 6, 29, 194, 199, 201, 205–6
World Meteorological Organization 58, 234
World Weather Attribution 14, 130

Yorkshire 222–4